MATRIX SPACES
AND
SCHUR MULTIPLIERS
Matriceal Harmonic Analysis

MATRIX SPACES
AND
SCHUR MULTIPLIERS
Matriceal Harmonic Analysis

Lars-Erik Persson

Luleå University of Technology, Sweden & Narvik University College, Norway

Nicolae Popa

"Simion Stoilov" Institute of Mathematics, Romanian Academy, Romania & Technical University "Petrol si Gaze", Romania

 **World Scientific**

NEW JERSEY • LONDON • SINGAPORE • BEIJING • SHANGHAI • HONG KONG • TAIPEI • CHENNAI

Published by

World Scientific Publishing Co. Pte. Ltd.

5 Toh Tuck Link, Singapore 596224

USA office: 27 Warren Street, Suite 401-402, Hackensack, NJ 07601

UK office: 57 Shelton Street, Covent Garden, London WC2H 9HE

Library of Congress Cataloging-in-Publication Data
Persson, Lars-Erik, 1944–   author.
   Matrix spaces and Schur multipliers : matriceal harmonic analysis / by Lars-Erik Persson (Luleå
University of Technology, Sweden & Narvik University College, Norway) & Nicolae Popa ("Simion
Stoilov" Institute of Mathematics, Romanian Academy, Romania & Technical University "Petrol si
Gaze", Romania).
      pages cm
   Includes bibliographical references and index.
   ISBN 978-9814546775 (alk. paper)
   1. Matrices. 2. Algebraic spaces. 3. Schur multiplier. I. Popa, Nicolae, author. II. Title.
   QA188.P43 2014
   512.9'434--dc23
                                                                    2013037182

British Library Cataloguing-in-Publication Data
A catalogue record for this book is available from the British Library.

Printed in Singapore

To Irina, Andrei, Manuela and Alexandra

Preface

In the last two centuries the Fourier analysis, known also as harmonic analysis, experienced a strong development. Roughly speaking it consists mainly in the study of properties of periodical functions connected to their Fourier coefficients. As examples of such properties we may consider the beautiful Fejer's theory [31] about the convergence of the Fourier series to a function with respect to its Cesaro means, or the well-known Fejér-Hardy-Littlewood inequality [96].

On the other hand many mathematicians have observed that there is a linear bijective correspondence between the periodical functions f on the torus $\mathbb{T}$ and its corresponding Toeplitz matrix, that is the infinite matrix having the nth Fourier coefficient of f on the nth diagonal submatrix parallel to the main diagonal, numbered by 0. (See for instance [94], [11].) For instance we began to think about this topic after reading the paper of J. Arazy [1], where a short remark about the analogy between Fourier coefficients and diagonal submatrices was made. In fact, it appears that periodical functions on the torus are particular cases of infinite matrices and it is tempting to develop a matrix version of the classical harmonic analysis, where instead of nth Fourier coefficient of a periodical function you have to consider the nth diagonal of a given infinite matrix.

The present volume is dedicated to this goal, namely to formulate and prove some statements in this new matrix version of classical harmonic analysis in terms of "diagonal" submatrices of an infinite matrix, which are analogous of the well-known statements in harmonic analysis. The current knowledge is presented in a unified way and also some new results are included to complete the picture to a fairly nice theory we hereby call *matriceal harmonic analysis*.

Now we briefly describe some of the most motivating results of this book.

After an introduction and a presentation of principal notions, in Chapter 2 we present some results from the master dissertation of Victor Lie (see [50]). The main idea of this work is to consider an infinite matrix as a sequence of functions $(\mathcal{L}_k)_{k\geq 1}$ and to exploit this interpretation in order to obtain a useful formula for the operator norm of an infinite matrix, namely:

$$||A||_{B(\ell_2)} = \sup_{||h||_2 \leq 1} ||V_{\mathcal{L}_B,h}||_\infty,$$

where

$$V_{\mathcal{L}_B,h}(x) = \left(\sum_{k=1}^\infty |(\mathcal{L}_k * h)(x)|^2\right)^{1/2}, \quad \forall x \in [0,1] \text{ and } \forall h \in H_0^2([0,1]).$$

By using this formula we give new proofs of some classical results of infinite matrix theory. For instance, it is possible to prove, with the same method, some scattered facts like:

- *For a Toeplitz matrix A, $A \in B(\ell_2)$ if and only if $f_A \in L^\infty([0,1])$* (see Theorem 3.2),
- (**Bennett's Theorem**) *For a Toeplitz matrix A, $A \in M(\ell_2)$ if and only f_A is a bounded Borel measure on $[0,1]$* (see Theorem 3.3),
- (**Nehari's Theorem**) *For a Hankel matrix A, $A \in B(\ell_2)$ if and only if $g_A \in BMO$* (see Theorem 3.5),
- (**Theorem of Kwapien-Pelczynski**) *If P_n is the main triangle projection of order n, then*

$$\sup_{||A||_{B(\ell_2)} \leq 1} ||P_n(A)||_{B(\ell_2)} = \mathcal{O}(\log n) \quad n \to \infty$$

(see Theorem 3.8).

In Chapter 3 we give a matrix version of Fejer's theory, introducing a subclass of infinite matrices representing bounded linear operators on ℓ_2, namely the class of *continuous matrices* $C(\ell_2)$.

This class consists of those infinite matrices, which are approximable in the operator norm by matrices of finitely band-type. In particular, we present the results in [17], which were extended later in [48] and [46].

Just as one possible application of the matrix version of Fourier analysis we mention the approximation of an infinite matrix in various ways. For instance the following natural question arise:

Which infinite matrices can be approximated in the operator norm by its finite-type band submatrices?

In particular, an answer to this question, which extends the well-known Jordan's theorem, is given in this chapter.

Another type of approximation of infinite matrices by a special type of matriceal polynomials is also given in Chapter 3 and it extends a well-known theorem of A. Haar from 1910 (see Theorem C on page 49):

- *Let $A = (a_k^l)_{l \geq 1, \, k \in \mathbb{Z}}$ be a matrix belonging to $C(\ell_2)$ such that all sequences $\mathbf{a}_k \overset{def}{=} (a_k^l)_{l \geq 1}$, $k \in \mathbb{Z}$, located on the kth diagonal, belong to the class ms.*
  Then, for any $\epsilon > 0$ there is an $n_\epsilon \in \mathbb{N}^*$ and sequences $\alpha_k \in ms$, $k \in \{0, \dots, n-1\}$, such that

$$\left\| A - \sum_{k=0}^{n-1} \alpha_k \odot H_k \right\|_{B(\ell_2)} < \epsilon.$$

Here ms is a special linear subspace of the space of all bounded sequences ℓ_∞, and $\odot$ means a product between two matrices, which extends the usual product of a scalar and a function.

Next we mention that in 1983 A. Shields [84] stated and proved the following beautiful theorem:

- *Let $M \in C_1$ have upper triangular form with respect to the orthonormal basis $\{e_n\}$ $(n = 1, 2, \dots)$ of ℓ_2. Then*

$$\sum_{k=1}^{\infty} \sum_{j=1}^{k} \frac{|M(j,k)|}{1+k-j} \leq \pi \|M\|_{T_1},$$

  with equality only when $M = 0$.

Here T_1 means the subspace of all trace class infinite upper triangular matrices.

The above result is analogous to the well-known Hardy-Littlewood-Fejer inequality.

We present and discuss this theorem and some other related results in Chapter 4.

For instance we present a new characterization of the elements of the space T_1 with respect to the sequence of a special linear combination of their diagonals, namely:

- Let $A \in B(\ell_2)$ be an upper triangular matrix. Then the following assertions are equivalent:

$$a)\ A \in T_1;$$

$$b)\ \sup_n \frac{1}{a_n} \sum_{j=0}^{n} \frac{1}{j+1} ||s_j A|| < \infty;$$

$$c)\ \sup_n ||P_n A|| < \infty.$$

Here

$$P_n A = \frac{1}{a_n} \sum_{j=0}^{n} \frac{1}{j+1} s_j A, \ \text{where} \ a_n = \sum_{j=0}^{n} \frac{1}{j+1} \ (n = 0, 1, 2, \dots)$$

and $s_j A = \sum_{k=0}^{j} A_k$.

It is important to mention that in Chapter 4 and also in the sequel, often the obtained matriceal results are only analogous to but do not extend the known results from harmonic analysis. This happens because most of the results from these chapters refer to matrices connected to Schatten classes of matrices and consequently their proofs cannot be applied to Toeplitz matrices.

Another topic which was touched some years ago is the study of matrix version of Hankel operators. Some especially interesting results about these Hankel operators were obtained by S. Power in [77]. For instance he showed there the following matrix version of Nehari's theorem:

- Let Φ be an infinite matrix such that $\Phi A \in C_2$ for all finite band-type matrices A.

  Then the following statements are equivalent:

  (a) H_Φ is a bounded linear operator on T_2.

  (b) There is $\Psi \in B(\ell_2)$ such that $\Psi_k = \Phi_k$ for all $k < 0$.

  (c) $P_- \Phi \in BMO_F(\ell_2)$.

See Chapter 5 for all unexplained notions and notations in this theorem.

In Chapter 5 we give a different and in our opinion more natural proof of this result and investigate this topic further. For example, we derive a sufficient condition in order that a matrix version of a Hankel operator to be nuclear. For a particular class of its symbol this condition is even necessary.

A class of Banach spaces of analytic functions, which has received great attention in the last two decades is the class of Bergman spaces. See e.g. [94], [28] and the references given there. We investigate some properties of the class of Bergman-Schatten spaces in Chapter 6. For instance we present and discuss some inequalities valid in Bergman-Schatten spaces (see [76]) e.g. the following:

- (**Hausdorff-Young Theorem**) *For* $1 \leq p \leq \infty$, *let* q *be the conjugate index, i.e.* $\frac{1}{p} + \frac{1}{q} = 1$ *(for* $p = 1$ *we have* $q = \infty$).

  (i) *If* $1 \leq p \leq 2$, *then* $A \in T_p$ *implies that* $\left(\sum_{n=0}^{\infty} ||A_n||_{T_p}^q \right)^{1/q} \leq ||A||_{T_p}$.

  (ii) *If* $2 \leq p \leq \infty$, *then* $\{||A_n||_{T_p}\} \in \ell_q$ *implies that* $||A||_{T_p} \leq \left(\sum_{n=0}^{\infty} ||A_n||_{T_p}^q \right)^{1/q}$.

Another important result, which appears in Chapter 6 is a matriceal analogue of a result obtained by Mateljevic and Pavlovic [61] in 1984, (see [62]):

- *Let* A *be an upper triangular matrix. Then* $A \in L_a^1(D, \ell_2)$ *if and only if*

$$\sum_{n=1}^{\infty} \frac{||\sigma_n(A)||_{C_1}}{(n+1)^2} < \infty.$$

Moreover, a matrix version of the usual Bloch space is introduced in Chapter 7. See [72]. The interest of this space consists mainly in the fact that it satisfies the equality

$$\mathcal{B}(D, \ell_2) = \left(H^1(\ell_2), BMOA(\ell_2) \right),$$

where $\mathcal{B}(D, \ell_2)$ means the matriceal Bloch space, $H^1(\ell_2)$, $BMOA(\ell_2)$ stand for the matrix version of the Hardy space H^1, respectively, for the matrix version of the space $BMOA$, and (X, Y) is the space of all Schur multipliers between spaces of infinite matrices X and Y.

This equation is the matrix analogue of a result valid for Fourier multipliers of periodical functions proved in 1990 by Mateljevic and Pavlovic [60].

Moreover, in 1995 O. Blasco [12] proved that a vector valued version of the equality of Mateljevic and Pavlovic is valid, generally speaking, only for functions with values in Hilbert spaces.

Consequently, this version of the above equality deserves some special attention. The equality is proved in Chapter 8. This chapter is dedicated to a very important tool in the theory, namely Schur multipliers, which represents the matrix version of classical Fourier multipliers.

We also present, prove and discuss some other results concerning Schur multipliers between Banach spaces of infinite upper triangular matrices. We mention just the following matrix version of a well-known result of Paley:

- If $A = \sum A_n \in H^1(\ell_2)$, then $P(A) := \sum_{k=1}^{\infty} A_{2^k} \in H^2(\ell_2)$.

Here $H^1(\ell_2)$ is a matrix version of Hardy space introduced in Chapter 4 and $H^2(\ell_2)$ is the space of all upper triangular Hilbert-Schmidt matrices.

Aknowledgement The second author was partially supported by the CNCSIS grant ID-PCE 1905/2008. Moreover, we are both grateful to Luleå University of Technology for financial support for research visits to be able to finalize this book. We are also very grateful to Dr. Niklas Grip for helping and supporting us in both professional and practical ways.

Special thanks are due to our colleagues Dr. A. N. Marcoci and Dr. L. G. Marcoci who read carefully a preliminary version of this book and made valuable suggestions and remarks.

Finally, we emphasize that writing this volume had not been possible without the existence of an atmosphere favorable to scientific activity, the atmosphere existing in the Institute of Mathematics "Simion Stoilow" of the Romanian Academy and at the Department of Mathematics at Luleå University of Technology.

The cover images of the midnight sun were taken by Elin Persson from the north harbour of Luleå, Sweden, at midnight of 16th June 2013. These images reflect the empowering senses the authors felt when this special book was finalized under the skies of this famous light.

Luleå, June 2013, under the influence of the magic midnight sun atmosphere close to the Polar Circle.

Lars-Erik Persson and Nicolae Popa

Contents

Preface   vii

1. Introduction   1

   1.1   Preliminary notions and notations   1
         1.1.1   Infinite matrices   1
         1.1.2   Analytic functions on disk   4
         1.1.3   Miscellaneous   5
         1.1.4   The Bergman metric   7

2. Integral operators in infinite matrix theory   9

   2.1   Periodical integral operators   9
   2.2   Nonperiodical integral operators   17
   2.3   Some applications of integral operators in the classical the-
         ory of infinite matrices   18
         2.3.1   The characterization of Toeplitz matrices   18
         2.3.2   The characterization of Hankel matrices   24
         2.3.3   The main triangle projection   27
         2.3.4 $B(\ell_2)$ is a Banach algebra under the Schur product 30

3. Matrix versions of spaces of periodical functions   33

   3.1   Preliminaries   34
   3.2   Some properties of the space $C(\ell_2)$ 34
   3.3   Another characterization of the space $C(\ell_2)$ and
         related results   36
   3.4   A matrix version for functions of bounded variation . . .   41

3.5   Approximation of infinite matrices by matriceal Haar poly-
        nomials   44
        3.5.1   Introduction   45
        3.5.2   About the space ms   50
        3.5.3   Extension of Haar's theorem   56
3.6   Lipschitz spaces of matrices; a characterization   61

4. Matrix versions of Hardy spaces   65

    4.1   First properties of matriceal Hardy space   65
    4.2   Hardy-Schatten spaces   69
    4.3   An analogue of the Hardy inequality in T_1   75
    4.4   The Hardy inequality for matrix-valued analytic functions   79
            4.4.1   Vector-valued Hardy spaces H_X^p   79
            4.4.2 $(H^p - \ell_q)$-multipliers and induced operators for
                     vector-valued functions   80
    4.5   A characterization of the space T_1   97
    4.6   An extension of Shields's inequality   101

5. The matrix version of $BMOA$ 109

    5.1   First properties of $BMOA(\ell_2)$ space 109
    5.2   Another matrix version of BMO and matriceal Hankel op-
            erators 111
    5.3   Nuclear Hankel operators and the space $\mathcal{M}^{1,2}$ 119

6. Matrix version of Bergman spaces   121

    6.1   Schatten class version of Bergman spaces   121
    6.2   Some inequalities in Bergman-Schatten classes   132
    6.3   A characterization of the Bergman-Schatten space   136
    6.4   Usual multipliers in Bergman-Schatten spaces   141

7. A matrix version of Bloch spaces   149

    7.1   Elementary properties of Bloch matrices   149
    7.2   Matrix version of little Bloch space   161

8. Schur multipliers on analytic matrix spaces   175

Bibliography   185

Index   191

Chapter 1

Introduction

1.1 Preliminary notions and notations

In this section we collect some notions and facts of the theory of infinite matrices, the theory of analytic functions on the disk and the circle, of vector-valued integration theory and of geometry of the disk etc.

1.1.1 *Infinite matrices*

For an infinite matrix $A = (a_{ij})$, and an integer k we denote by A_k the matrix whose entries $a'_{i,j}$ are given by

$$a'_{i,j} = \begin{cases} a_{i,j} & \text{if } j - i = k, \\ 0 & \text{otherwise} \end{cases}.$$

Then A_k will be called *the kth-diagonal matrix associated to A.*

Sometimes we use also the notation $a(i, j)$ for the entries of the matrix A.

An important notion in the theory of matrices is the *Schur product*. Let $A = (a_{ij})_{i,j}$ and $B = (b_{ij})_{i,j}$ be two infinite matrices. Then the *Schur product* $C = (c_{ij})_{i,j}$ *of A and B, denoted by* $A * B$, has the entries $c_{ij} = a_{ij}b_{ij}$ for all $i, j \in \mathbb{N}$.

An infinite matrix A such that $A * B \in Y$ for all $B \in X$, where X, Y are Banach spaces of infinite matrices, is called a *Schur multiplier from X into Y*, and the space of all Schur multipliers from X into Y, endowed with the natural norm

$$||A||_{(X,Y)} = \sup_{||B||_X \leq 1} ||A * B||_Y$$

is denoted by (X, Y).

In the case $X = Y = B(\ell_2)$, where $B(\ell_2)$ is the space of all linear and bounded operators on ℓ_2, the space (X, Y) is denoted by $M(\ell_2)$ (an explanation of this notation is given later in this section) and a matrix $A \in M(\ell_2)$ is simply called a *Schur multiplier*.

We consider on the interval $[0, 1)$ the Lebesgue measurable infinite matrix valued functions $A(r)$. These functions may be regarded as infinite matrix-valued functions defined on the unit disk D using the correspondence $A(r) \rightarrow f_A(r, t) = \sum_{k=-\infty}^{\infty} A_k(r) e^{ikt}$, where $A_k(r)$ is the kth-diagonal of the matrix $A(r)$, the preceding sum is a formal one and t belongs to the torus $\mathbb{T}$.

We may consider $f_A(r, t)$, or $f_A(z)$, with $z = re^{it}$, as a matrix valued function, or distribution, or just a formal series.

Such a matrix $A(r)$ is called *an analytic matrix* if there exists an upper triangular infinite matrix A such that, for all $r \in [0, 1)$, we have $A_k(r) = A_k r^k$, for all $k \in \mathbb{Z}$.

In what follows we identify the analytic matrices $A(r)$ with their corresponding upper triangular matrices A and call the latter also as *analytic matrices*.

A special class of infinite matrices is considered often in this book, namely the class of *Toeplitz matrices*.

Let $A = (a_{ij})_{i,j \geq 1}$ be an infinite matrix. If there is a sequence of complex numbers $(a_k)_{k=-\infty}^{+\infty}$, such that $a_{ij} = a_{j-i}$ for all $i, j \in \mathbb{N}$, then A is called a *Toeplitz matrix*.

For simplicity we can identify a Toeplitz matrix with $A = (a_k)_{k=-\infty}^{+\infty}$, and the class of all Toeplitz matrices is denoted by $\mathcal{T}$.

G. Bennett proved in 1977 the following interesting result (see Theorem 8.1-[11]) about Schur multipliers:

Bennett's Theorem *The Toeplitz matrix* $M = (c_{j-k})_{j,k}$, *where* $(c_n)_{n \in \mathbb{Z}}$ *is a sequence of complex numbers, is a Schur multiplier if, and only if, there exists a (bounded, complex, Borel) measure μ on (the circle group)* $\mathbb{T}$ *with*

$$\widehat{\mu}(n) = c_n \quad for\ n = 0, \pm 1, \pm 2, \ldots.$$

Moreover, we then have

$$||M||_{M(\ell_2)} = ||\mu||.$$

Bennett's theorem justifies the notation $M(\ell_2)$ since for a Toeplitz matrix the notions of Schur multiplier and Borel measure on the torus coincide.

In the sequel we give some results about compact operators on the Hilbert space ℓ_2. (See for instance [94].)

For example the following decomposition formula is known.

Schmidt Theorem *If T is a self-adjoint compact operator on a Hilbert space H, then there exists a sequence of real numbers $\{\lambda_n\}$ tending to 0 and there also exists an orthonormal set $\{e_n\}$ in H such that*

$$Tx = \sum_{n=1}^{\infty} \lambda_n(x, e_n)e_n$$

for all $x \in H$.

If the operator T is compact, but not necessarily self-adjoint, then we first consider the polar decomposition $T = V|T|$, where $|T| = (T^*T)^{1/2}$ is positive (and hence self-adjoint) and compact. By the above theorem, there is an orthonormal set $\{e_n\}$ in H such that

$$|T|x = \sum_n \lambda_n(x, e_n)e_n, \quad x \in H,$$

where $\{\lambda_n\}$ is a nonincreasing sequence of nonnegative numbers tending to 0. Let $\sigma_n = Ve_n$ for each n; then $\{\sigma_n\}$ is still an orthonormal set and we have that

$$Tx = \sum_n \lambda_n(x, e_n)\sigma_n, \quad x \in H.$$

This is called *the canonical decomposition of a compact operator T*. The non-increasingly arranged sequence $\{\lambda_n\}$ is called the sequence of *singular values of T*. The number λ_n is called the nth singular value of T.

Now we introduce the *Schatten class operators*.

Given $0 < p < \infty$ and a separable Hilbert space H, we define the Schatten p-class of H, denoted by $C_p(H)$ or simply C_p, to be the space of all compact operators T on H with its singular value sequence $\{\lambda_n\}$ belonging to ℓ_p (the *p-summable sequence space*). We are mainly concerned with the range $1 \leq p < \infty$. In this case, C_p is a Banach space with the norm

$$||T||_p = \left[\sum_n |\lambda_n|^p\right]^{1/p}.$$

We recall that we can identify the operators with its corresponding matrices; so we may consider C_p as being spaces of matrices.

C_1 is also called the *trace class* of H, and C_2 is usually called the *Hilbert-Schmidt class*.

1.1.2   *Analytic functions on disk*

In this subsection we introduce the definitions of some important spaces of analytic functions on the disk.

First of all we consider the classical *Hardy space* of functions on the disk.

Let $0 < p \leq \infty$. We say that an analytic function $f : D \to \mathbb{C}$ belongs to the Hardy space H^p, for $1 \leq p < \infty$ if and only if $||f||_{H^p} := \sup_{0 \leq r < 1} \left(\frac{1}{2\pi} \int_0^{2\pi} |f(re^{i\theta})|^p d\theta \right)^{1/p} < \infty$ and $f \in H^\infty$ if and only if $||f||_{H^\infty} := \sup_{0 \leq \theta \leq 2\pi;\, 0 \leq r < 1} |f(re^{i\theta})| < \infty$.

The Hardy space was intensely studied in the last forty years. Duren's book [23] and Garnett's book [33] are frequently cited as references to H^1.

We mention only that Hardy and Littlewood proved in [39] the following inequality for functions $f \in H^1$:

$$(HL) \quad \int_0^1 (1 - r) \left(\int_0^1 |f'(re^{2\pi i\theta})| d\theta \right)^2 dr \leq C||f||_{H^1}^2,$$

where $C > 0$ is a constant independent of the choice of f.

Another analytic function space studied in connection with Hardy space is the space of all analytic functions of bounded mean oscillation denoted by $BMOA$.

One of the equivalent definitions of this space is as follows (see [33]): $BMOA$ coincides with the space of all analytic functions f on the disk such that the following norm is finite:

$$||f||_* := \sup_{\lambda \in D} \left\{ \int \int_D |f'(z)|^2 (1 - |z|^2)(1 - |\lambda|^2)|1 - \bar{\lambda}z|^{-2} dx dy \right\}^{1/2}.$$

Now we recall the definition of the classical *Bloch space* for functions on the disk.

The Bloch space $\mathcal{B}$ is the space of all analytic functions $f : D \to \mathbb{C}$ such that

$$||A||_\mathcal{B} := \sup_{z \in D} (1 - |z|^2)|f'(z)| + |f(0)| < \infty.$$

Another interesting space of analytic functions on the disk is the so-called *little Bloch space*.

The little Bloch space of D, denoted by $\mathcal{B}_0$, is the closed subspace of $\mathcal{B}$ consisting of functions f with $(1 - |z|^2)f'(z) \to 0$ $(|z| \to 1^-)$.

1.1.3 *Miscellaneous*

We also need some notions of vector-valued integration theory.

We say that a function $f : D \to B(\ell_2)$, is w^*-*measurable* if $A \circ f$ is a Lebesgue measurable function on D for every $A \in C_1$, where C_1 is the *Schatten class of all operators with trace*, and A is considered as a functional on $B(\ell_2)$.

The function $f : D \to B(\ell_2)$ is *strongly measurable* if it is a norm limit of a sequence of simple functions. (See [30] for more details about vector-valued measurability.)

Then we have the following particular case of Proposition 8.15.3-[30]:

Proposition 1.1. *Let $f : D \to B(\ell_2)$ be a w^*-mesurable function. Then the function $\|f\| : z \to \|f(z)\|_{B(\ell_2)}$ is measurable.*

Moreover, we state without proof the following theorem, which is a particular case of Theorem 8.18.2 [30] for $E = C_1$:

Theorem 1.2. *The topological dual of $L^1(D, \ell_2)$ may be identified with $L^\infty(D, \ell_2)$ by the duality bilinear map:*

$$< A(\cdot), B(\cdot) > := \int_0^1 tr\ (A(s)[B(s)]^*) 2s ds,$$

where $A(\cdot) \in L^\infty(D, \ell_2)$, $B(\cdot) \in L^1(D, \ell_2)$.

Here $L^1(D, \ell_2)$ and $L^\infty(D, \ell_2)$ are defined in Section 6.1 as particular cases of Banach spaces of vector-valued functions.

We also need the following well-known lemma (see for instance [94]):

Lemma 1.3. *Let $z \in D$, $c \in \mathbb{R}$, $t > -1$, and*

$$I_{c,t}(z) = \int_D \frac{(1 - |w|^2)^t}{|1 - z\overline{w}|^{2+t+c}} dA(w).$$

Then we have that

(1) if $c < 0$, then $I_{c,t}(z)$ is bounded in z;
(2) if $c > 0$, then

$$I_{c,t}(z) \sim \frac{1}{(1 - |z|^2)^c} \quad (|z| \to 1^-);$$

(3) if $c = 0$, then

$$I_{0,t}(z) \sim \log \frac{1}{1 - |z|^2} \quad (|z| \to 1^-).$$

Proof.   Since $t > -1$, the integral $I_{c,t}$ is defined for all $z \in D$. Let $\lambda = \frac{1}{2}(2 + t + c)$. If λ is zero or a negative integer, then clearly $c < 0$ and $I_{c,t}(z)$ is bounded. If λ is not zero or a negative integer, then

$$\frac{1}{(1 - z\overline{w})^\lambda} = \sum_{n=0}^{\infty} \frac{\Gamma(n + \lambda)}{n!\Gamma(\lambda)} z^n \overline{w}^n$$

and the rotation invariance of $(1 - |w|^2)^t dA(w)$ shows that

$$\int_D \frac{(1 - |w|^2)^t}{|1 - z\overline{w}|^{2\lambda}} dA(w) = \sum_{n=0}^{\infty} \frac{\Gamma(n + \lambda)^2}{(n!)^2 \Gamma(\lambda)^2} |z|^{2n} \int_D (1 - |w|^2)^t |w|^{2n} dA(w)$$

$$= \sum_{n=0}^{\infty} \frac{\Gamma(n + \lambda)^2}{(n!)^2 \Gamma(\lambda)^2} |z|^{2n} \int_0^1 (1 - r)^t r^n dr$$

$$= \sum_{n=0}^{\infty} \frac{\Gamma(n + \lambda)^2}{(n!)^2 \Gamma(\lambda)^2} \frac{\Gamma(t + 1)\Gamma(n + 1)}{\Gamma(n + t + 2)} |z|^{2n}$$

$$= \frac{\Gamma(t + 1)}{\Gamma(\lambda)^2} \sum_{n=0}^{\infty} \frac{\Gamma(n + \lambda)^2}{n!\Gamma(n + t + 2)} |z|^{2n}.$$

By Stirling's formula,

$$\frac{\Gamma(n + \lambda)^2}{n!\Gamma(n + t + 2)} \sim n^{c-1} \quad (n \to \infty).$$

Clearly $\sum_{n=1}^{\infty} n^{c-1}|z|^{2n}$ is bounded in z for $c < 0$. If $c = 0$, then

$$\sum_{n=1}^{\infty} \frac{|z|^{2n}}{n} = \log \frac{1}{1 - |z|^2}.$$

If $c > 0$, then

$$n^{c-1}|z|^{2n} \sim \frac{1}{(1 - |z|^2)^c}$$

since

$$\frac{1}{(1 - |z|^2)^c} = \sum_{n=0}^{\infty} \frac{\Gamma(n + c)}{n!\Gamma(c)} |z|^{2n}$$

and $\Gamma(n + c)/n! \sim n^{c-1}$ by Stirling's formula. This completes the proof of the lemma. $\square$

1.1.4 *The Bergman metric*

In this section we recall some geometric facts about the disk. More specifically we consider the Bergman metric on D, which will be useful in the study of matriceal Bloch space.

The pseudo-hyperbolic metric. Recall that for any $z \in D$, φ_z is the Moebius transformation of D, which interchanges the origin and z, namely

$$\phi_z(w) = \frac{z - w}{1 - \bar{z}w}, \quad w \in D.$$

The pseudo-hyperbolic distance on D is defined by

$$\rho(z, w) = |\varphi_z(w)| = \left| \frac{z - w}{1 - \bar{z}w} \right|, \quad z, w \in D.$$

An important property of the pseudo-hyperbolic distance is that it is Moebius invariant, that is,

$$\rho(\varphi(z), \varphi(w)) = \rho(z, w)$$

for all $\varphi \in Aut(D)$, the Moebius group of D, and all $z, w \in D$.

The Bergman metric.

The Bergman metric on D is given by

$$\beta(z, w) = \frac{1}{2} \log \frac{1 + \rho(z, w)}{1 - \rho(z, w)}, \quad z, w \in D.$$

The Bergman metric is also Moebius invariant:

$$\beta(\varphi(z), \varphi(w)) = \beta(z, w)$$

for all $\varphi \in Aut(D)$ and all $z, w \in D$.

Notes

Most of the information in this chapter is classical but not so easy to find collected in this form elsewhere. The Schur (or Hadamard) product is well known to specialists but there are few monographs which treat this matter. The authors know only the book of G. Pisier [81]. Bennett's Theorem is also considered as part of folklore (see for example [8]). For an accessible proof see [11] and [8].

In the next chapter we present another proof of this important result (see Theorem 2.14).

More about Schmidt Theorem and Schatten class operators can be found in many excellent books, for instance [34].

Analytic function on the disk are treated also in many books as [68], [33] and [23].

We pay special attention to different spaces of analytic functions on the disk as Hardy space H^1, which is intensively studied in [23] and [33]. More about the space of analytic functions of bounded mean oscillation $BMOA$ may be found in [33].

A very interesting space is also the Bloch space of analytic functions $\mathcal{B}$. A classical reference is [3].

Proposition 1.1, Theorem 1.2 and related facts about vector-valued integration theory are mainly taken from [30]. Lemma 1.3 may be found in [94]. Moreover, the subsection dedicated to the Bergmann metric is taken from [94].

Chapter 2

Integral operators in infinite matrix theory

The content of the present chapter is taken from the master dissertation of V. Lie at University of Bucharest under the supervision of the second author (see [50]).

We start defining and discussing some important devices we need in what follows in the study of infinite matrices.

The main idea is to consider an infinite matrix as a sequence of functions (resp. distributions). This new point of view has the advantage to use the more refined results from function theory in the theory of infinite matrices.

For instance in the first section we define the important notions of square function and matriceal operator associated to a matrix. Using these notions we prove the first main result, namely Theorem 2.8.

In the second section the central result is the non-periodical analogue of Theorem 2.8 (see Proposition 2.2).

2.1 Periodical integral operators

Let the matrix

$$
B = \begin{pmatrix}
b_{11} & b_{12} & b_{13} & \cdots \\
b_{21} & b_{22} & b_{23} & \cdots \\
\vdots & \vdots & \vdots & & \cdots \\
b_{n1} & b_{n2} & b_{n3} & \cdots \\
\vdots & \vdots & \vdots & & \cdots
\end{pmatrix} \in B(\ell_2).
$$

Since $B \in B(\ell_2)$ it follows that, for all $k \in \mathbb{N}$, the sequence $(b_{kj})_{j\geq 1} \in$

$\ell_2(\mathbb{N})$. Therefore we can define the functions:

$$\mathcal{L}_k(B)(t) = \sum_{j=1}^{\infty} b_{kj} e^{2\pi ijt} \in H_0^2([0,1]),$$

and

$$\mathcal{L}_k^*(B)(t) = \sum_{j=1}^{\infty} b_{kj} e^{2\pi ijt} \cdot e^{-2\pi ikt} = \mathcal{L}_k(B)(t) e^{-2\pi ikt}.$$

Consequently, to each row k in the matrix B it corresponds a unique function from the Hardy space, $H_0^2([0,1])$, of all analytic functions $h(t) = \sum_{k=1}^{\infty} x_k e^{2\pi ikt}$, $\forall t \in [0,1]$. This function is denoted by $\mathcal{L}_k(B)$, and $\widehat{\mathcal{L}_k(B)}(j) = b_{kj}$, for all $k, j \geq 1$.

For brevity we denote in what follows $\mathcal{L}_k(B)$ simple by $\mathcal{L}_k$, and $\mathcal{L}_k^*(B)$ by $\mathcal{L}_k^*$.

Thus the matrix B can be written as follows:

$$B = \begin{pmatrix} \int_0^1 \mathcal{L}_1(t) e^{-2\pi it} dt & \int_0^1 \mathcal{L}_1(t) e^{-4\pi it} dt & \dots \\ \int_0^1 \mathcal{L}_2(t) e^{-2\pi it} dt & \int_0^1 \mathcal{L}_2(t) e^{-4\pi it} dt & \dots \\ \vdots & \vdots & \dots \\ \int_0^1 \mathcal{L}_n(t) e^{-2\pi it} dt & \int_0^1 \mathcal{L}_n(t) e^{-4\pi it} dt & \dots \\ \vdots & \vdots & \dots \end{pmatrix},$$

or

$$B =$$

$$\begin{pmatrix} \int_0^1 \mathcal{L}_1^*(t) dt & \int_0^1 \mathcal{L}_1^*(t) e^{-2\pi it} dt & \int_0^1 \mathcal{L}_1^*(t) e^{-4\pi it} dt & \dots \\ \int_0^1 \mathcal{L}_2^*(t) e^{2\pi it} dt & \int_0^1 \mathcal{L}_2^*(t) dt & \int_0^1 \mathcal{L}_2^*(t) e^{-2\pi it} dt & \dots \\ \vdots & \vdots & \dots & \dots \\ \int_0^1 \mathcal{L}_n^*(t) e^{2\pi i(n-1)t} dt & \int_0^1 \mathcal{L}_n^*(t) e^{2\pi i(n-2)t} dt & \int_0^1 \mathcal{L}_n^*(t) e^{2\pi i(n-3)t} dt & \dots \\ \vdots & \vdots & \vdots & \dots \end{pmatrix}.$$

We use the both expressions of B in what follows.

If $x = (x_j)_{j \geq 1} \in \ell_2$, then

$$\|Bx\|_2^2 = \left| \int_0^1 \mathcal{L}_1(t) \left(x_1 e^{-2\pi it} + x_2 e^{-4\pi it} + \dots \right) dt \right|^2 + \dots$$

$$+ \left| \int_0^1 \mathcal{L}_n(t) \left(x_1 e^{-2\pi it} + x_2 e^{-4\pi it} + \dots \right) dt \right|^2 + \dots$$

Denote by $h(t)$ the sum $\sum_{j=1}^{\infty} x_j e^{2\pi ijt} \in H_0^2([0,1])$. Then

$$\|Bx\|_2^2 = \left| \int_0^1 \mathcal{L}_1(t) h(-t) dt \right|^2 + \dots + \left| \int_0^1 \mathcal{L}_n(t) h(-t) dt \right|^2 + \dots$$

$$= \sum_{k=1}^{\infty} \left| \int_0^1 \mathcal{L}_k(t) h(-t) dt \right|^2,$$

and, since $\mathcal{L}_k \in H_0^2([0,1])$ for all $k \geq 1$, we have that

$$\int_0^1 \mathcal{L}_k(t) h(-t) dt = \int_0^1 \mathcal{L}_k(t) g(-t) dt,$$

for all $g \in L^2([0,1])$ such that $\widehat{(h-g)}(n) = 0$ for all $n \geq 1$.
Thus,

$$\|B\|_{B(\ell_2)} = \sup_{\|x\|_2 \leq 1} \|Bx\|_2 = \sup_{\|h\|_{H^2} \leq 1} \left(\sum_{k=1}^{\infty} \left| \int_0^1 \mathcal{L}_k(t) h(-t) dt \right|^2 \right)^{1/2}$$

$$= \sup_{\|g\|_{L^2} \leq 1} \left(\sum_{k=1}^{\infty} \left| \int_0^1 \mathcal{L}_k(t) g(-t) dt \right|^2 \right)^{1/2}.$$

Consequently, the space $B(\ell_2)$ may be considered as a subspace of the space $\prod_{n=1}^{\infty} H_n$, where $H_n = H_0^2([0,1])$ for all $n \geq 1$. Moreover, if we denote by

$$\mathcal{L} := (\mathcal{L}_1, \mathcal{L}_2, \dots) \in \prod_{n=1}^{\infty} H_n,$$

and

$$\|\mathcal{L}\|_{H^2(\infty)} - \sup_{\|h\|_{H^2} \leq 1} \left(\sum_{k=1}^{\infty} \left| \int_0^1 \mathcal{L}_k(t) h(-t) dt \right|^2 \right)^{1/2},$$

then it follows

$$H_0^2(\infty) := \{ \mathcal{L} \in \prod_{n=1}^{\infty} H_n \mid \|\mathcal{L}\|_{H_0^2(\infty)} < \infty \},$$

and $\left(H_0^2(\infty), \| \cdot \|_{H_0^2(\infty)} \right)$ is a Banach space.

Moreover, the linear operator defined by

$$L : \left(B(\ell_2), || \, ||_{B(\ell_2)} \right) \to \left(H_0^2(\infty), || \, ||_{H_0^2(\infty)} \right)$$

$$L(B) = \mathcal{L}_B,$$

where

$$\mathcal{L}_B = (\mathcal{L}_1(B), \mathcal{L}_2(B), \dots)$$

is an isometry between $B(\ell_2)$ and $H_0^2(\infty)$.

In the sequel we introduce the periodical square function associated to a matrix.

Let $\mathcal{L} = (\mathcal{L}_1, \mathcal{L}_2, \dots) \in \prod_{n=1}^{\infty} H_n$ and $h \in H_0^2([0,1])$.

We define

$$V_{\mathcal{L},h} : [0,1] \to \overline{\mathbb{R}}_+$$

by:

$$V_{\mathcal{L},h}(x) = \left(\sum_{k=1}^{\infty} |(\mathcal{L}_k * h)(x)|^2 \right)^{1/2}.$$

Proposition 2.1. *Let $\mathcal{L} = (\mathcal{L}_1, \mathcal{L}_2, \dots)$ be a fixed element in $\prod_{n=1}^{\infty} H_n$. We have that*

$$\sup_{||h||_{H^2} \le 1} V_{\mathcal{L},h}(0) = \sup_{||h||_{H^2} \le 1} ||V_{\mathcal{L},h}(\cdot)||_\infty.$$

Proof. First, it is clear that

$$V_{\mathcal{L},h}(0) \le ||V_{\mathcal{L},h}||_\infty.$$

For the converse, observe that

$$V_{\mathcal{L},h_0}(x) \le \sup_{||h||_2 \le 1} V_{\mathcal{L},h}(x),$$

where $h_0 \in H_0^2([0,1])$ is fixed with $||h_0||_2 \le 1$.

We define the isometric operator

$$S_x(h) = h_x,$$

where $h_x(t) = h(x + t)$.

Then

$$\sup_{||h||_2 \le 1} V_{\mathcal{L},h}(x) = \sup_{||h||_2 \le 1} V_{\mathcal{L},h_x}(0) = \sup_{||h_x||_2 \le 1} V_{\mathcal{L},h_x}(0) = \sup_{||h||_2 \le 1} V_{\mathcal{L},h}(0),$$

and, hence,

$$V_{\mathcal{L},h_0}(x) \le \sup_{||h||_2 \le 1} V_{\mathcal{L},h}(0),$$

for all h_0 such that $||h_0||_2 \le 1$, and for all $x \in [0,1]$.

In other words

$$\sup_{||h||_2 \le 1} ||V_{\mathcal{L},h}||_\infty = \sup_{||h_0||_2 \le 1} ||V_{\mathcal{L},h_0}||_\infty \le \sup_{||h||_2 \le 1} V_{\mathcal{L},h}(0). \qquad \square$$

Remark 2.2. a) *By the previous discussion it follows that*

$$||\mathcal{L}||_{H_0^2(\infty)} = \sup_{||h||_{H^2} \leq 1} ||V_{\mathcal{L},h}||_\infty,$$

and, consequently,

$$H_0^2(\infty) = \{\mathcal{L} \in \prod_{n=1}^\infty H_n \mid \sup_{||h||_{H^2} \leq 1} ||V_{\mathcal{L},h}||_\infty < \infty\}.$$

b) It is clear from the definition that the value of $V_{\mathcal{L},h}$ in each point $x \in [0,1]$ has a definite meaning.

Proposition 2.3. *Let $\mathcal{L} = (\mathcal{L}_1, \mathcal{L}_2, \dots) \in \prod_{n=1}^\infty H_n$, and B the matrix canonically associated to this element in the sense defined previously. Then the following assertions are equivalent:*

i) $B \in B(\ell_2)$.

ii) $\mathcal{L}_B = \mathcal{L} \in H_0^2(\infty)$.

iii) For all $h \in H_0^2([0,1])$ we have $V_{\mathcal{L},h} \in C_s([0,1])$, where $C_s([0,1])$
$:= \{f : [0,1] \to \mathbb{C} \mid \exists (f_n)_{n\geq 1} \text{ such that } f_n : [0,1] \to \mathbb{C} \text{ are continuous functions, } (f_n)_{n\geq 1} \text{ is a bounded sequence in the sup-norm and } f_n(x) \to f(x) \text{ as } n \to \infty, \text{ for all } x \in [0,1]\}.$

Proof. The implication $i) \Rightarrow ii)$ follows from the definition of the operator L.

$i) \Rightarrow iii)$ We show that

$$V_{\mathcal{L},h}(x) \leq M||h||_2,$$

for $h \in H_0^2([0,1])$, $x \in [0,1]$, and $M := ||B||_{B(\ell_2)}$.

Of course, it is enough to prove that $V_{\mathcal{L},h}(x) \leq M$, for $||h||_2 \leq 1$.

If not, then there exist $x_0 \in [0,1]$, and $h_0 \in H_0^2([0,1])$, with $||h_0||_2 \leq 1$ such that

$$V_{\mathcal{L},h_0}(x_0) > M.$$

But, for $h_0(t) = \sum_{k=1}^\infty a_k e^{2\pi ikt}$, we have that

$$V_{\mathcal{L},h_0}(x_0) = (V_{\mathcal{L},h_0})_{x_0}(0) = ||By_0||_2,$$

where $y_0 = (y_0^1, y_0^2, \dots)$, and $y_0^k = a_k e^{2\pi ikx_0}$.

Since $||y_0||_2 = ||h_0||_{H^2}$ it follows that there exists y_0, with $||y_0||_2 \leq 1$ such that

$$||By_0||_2 > ||B||_{B(\ell_2)} = M,$$

which is a contradiction.

Therefore, for each $h \in H_0^2([0,1])$, and for each $x \in [0,1]$, we have

$$V_{\mathcal{L},h}(x) \le M||h||_2.$$

Fixing h it follows that there exists $C > 0$ such that, for all $x \in [0,1]$, we have that

$$V_{\mathcal{L},h}(x) < C.$$

Hence, the sum $\sum_{k=1}^{\infty} |(\mathcal{L}_k * h)(x)|^2$ converges absolutely and because $\mathcal{L}_k * h \in C([0,1])$, for all $k \ge 1$,

$$V_{\mathcal{L},h} \in C_s([0,1]).$$

$iii) \Rightarrow i)$ From the hypothesis we have that $V_{\mathcal{L},h}(0) < \infty$ for all $h \in H_0^2([0,1])$. This fact means, by using the correspondences $B \leftrightarrow (\mathcal{L}_k)_{k \ge 1} = \mathcal{L}$, and $x \leftrightarrow h$, that, for each $x \in \ell_{\mathbb{N}^*}$, we have $||Bx||_2 < \infty$.

In view of Banach Steinhaus Theorem it follows that

$$B \in B(\ell_2). \qquad \square$$

We extend in what follows some of the notions introduced previously.

Definition 2.4. *Let $B \in \mathcal{PM}(\ell_2) := \{A = (a_{ij})_{i,j \ge 1}$ such that $\exists C > 0$ with $|a_{ij}| \le C \forall i, j \ge 1\}$, and let $\mathcal{L}_B := \left(\mathcal{L}_k^B\right)_{k \ge 1}$ be its sequence of distributions. We call the matriceal distribution associated to the matrix B the expression*

$$\mathcal{L}_B \in \mathcal{D}'([0,1] \times [0,1])$$

given by the formula

$$\mathcal{L}_B(t, x) = \sum_{k=1}^{\infty} \mathcal{L}_k^B(t) e^{2\pi i k x}.$$

Remark 2.5. *i) If $\mathcal{L}_B = \left(\mathcal{L}_k^B\right)_{k \ge 1} \in \prod_{n=1}^{\infty} H_0^2([0,1])$, and $h \in H_0^2([0,1])$, then we observe that*

$$V_{\mathcal{L}_B,h}(u) = \left(\int_0^1 \left| \int_0^1 \mathcal{L}_B(t, x) h(u - t) dt \right|^2 dx \right)^{1/2}.$$

Hence,

$$B \in B(\ell_2) \Leftrightarrow \sup_{||h||_{H^2} \le 1} \left(\int_0^1 \left| \int_0^1 \mathcal{L}_B(t, x) \overline{h(t)} dt \right|^2 dx \right)^{1/2} < \infty.$$

ii) The notion of square function associated to a matrix may be extended from the class $B(\ell_2)$ to the class $\mathcal{PM}(\ell_2)$ as follows:

Let $\mathcal{P}([0,1])$ be the linear space of all analytic polynomials (i.e. trigonometrical polynomials having Fourier coefficients of nonpositives indices equal to zero). Then we define, $\forall\, h \in \mathcal{P}([0,1])$,

$$V_{\mathcal{L}_B,h} := \sup_{r \in \mathcal{P}([0,1]);\, ||r||_{H^2} \leq 1} \left| \int_0^1 \int_0^1 \mathcal{L}_B(t,x)h(u-t)\overline{r(x)}dtdx \right|.$$

Definition 2.6. *Let $B \in \mathcal{PM}(\ell_2)$ and let $\mathcal{L}_B$ denote the matriceal distribution associated to the matrix B. Then, the operator*

$$T_B : \mathcal{P}([0,1]) \times \mathcal{P}([0,1]) \to C([0,1] \times [0,1]),$$

defined by

$$T_B(r \otimes h)(u,v) = \int_0^1 \int_0^1 \mathcal{L}_B(t,x)r(u-x)h(v-t)dtdx,$$

is called the matriceal operator associated to B.

Proposition 2.7. *Let $B \in \mathcal{PM}(\ell_2)$. The following assertions are equivalent:*

i) $B \in B(\ell_2)$.

ii) There exists a continuous operator $\widetilde{T}_B : H_0^2([0,1])\overline{\otimes}H_0^2([0,1]) \to C_s([0,1] \times [0,1])$, such that

$$\widetilde{T}_B \big|_{\mathcal{P}([0,1])\otimes\mathcal{P}([0,1])} = T_B .$$

Proof. *i) $\Rightarrow$ ii)* Let $B \in B(\ell_2)$. Then it follows that, for all $h \in H_0^2([0,1])$, and for all $x \in [0,1]$, we have that

$$(a) \quad V_{\mathcal{L}_B,h}(x) \leq M||h||_2,$$

with $M = ||B||_{B(\ell_2)}$.

Let $r, h \in \mathcal{P}([0,1])$ be fixed polynomials. Then:

$$T_B(r \otimes h)(u,v) = \int_0^1 \int_0^1 \left(\sum_{k=1}^{\infty} \mathcal{L}_k^B(t)e^{2\pi ikx} \right) r(u-x)h(v-t)dxdt$$

$$= \sum_{k=1}^{\infty} \left(\int_0^1 \mathcal{L}_k^B(t)h(v-t)dt \right) \left(\int_0^1 r(u-x)e^{2\pi ikx}dx \right)$$

$$= \sum_{k=1}^{\infty} \left(\mathcal{L}_k^B * h \right)(v) \left(\int_0^1 r(x)e^{-2\pi ikx}dx \right) e^{2\pi iku}.$$

Hence,

$$|T_B(r \otimes h)(u,v)| \leq V_{\mathcal{L}_B,h}(v)||r||_2 \leq \text{(by (a))} \leq M||h||_2||r||_2,$$

and

$$||T_B(r \otimes h)||_\infty \leq ||B||_{B(\ell_2)}||r \otimes h||.$$

Since $\mathcal{P}([0,1])$ is dense in the norm $||\,||_2$, it follows that there exists a unique continuous extension $\widetilde{T}_B$ like in the statement of the proposition.

$ii) \Rightarrow i)$ Let $\widetilde{T}_B$ be as in the hypothesis. Then there exists $M > 0$ such that, for all $r, h \in \mathcal{P}([0,1])$, we have that

$$||T_B(r \otimes h)||_\infty \leq M||r \otimes h|| = M||r||_2||h||_2.$$

We fix $r, h \in \mathcal{P}([0,1])$. Then it follows that

$$||T_B(r \otimes h)||_\infty \geq |T_B(r \otimes h)(0,0)| = \left| \sum_{k=1}^\infty \left(\mathcal{L}_k^B * h \right)(0) \int_0^1 r(x) e^{-2\pi i k x} dx \right|,$$

and, therefore,

$$\left| \sum_{k=1}^\infty \left(\mathcal{L}_k^B * h \right)(0) \int_0^1 r(x) e^{-2\pi i k x} dx \right| \leq M||r||_2||h||_2.$$

Next we take the supremum over all $r \in \mathcal{P}([0,1])$ with $||r||_2 \leq 1$ and we get that

$$V_{\mathcal{L}_B,h}(0) \leq M||h||_2 \quad \forall h \in \mathcal{P}([0,1]).$$

Hence,

$$||B||_{B(\ell_2)} = \sup_{||h||_2 \leq 1} V_{\mathcal{L}_B,h}(0) \leq M < \infty,$$

i.e.

$$B \in B(\ell_2). \qquad \square$$

Consequently, we have:

Theorem 2.8. *Let $D \subset \mathcal{PM}(\ell_2)$. The following assertions are equivalent:*
i) $B \in B(\ell_2)$.
ii) $\mathcal{L}_B \in H_0^2(\infty)$.
iii) For all $h \in H_0^2([0,1])$, we have that $V_{\mathcal{L}_B,h} \in C_s([0,1])$.
iv) $\sup_{||h||_2 \leq 1} ||V_{\mathcal{L}_B,h}||_\infty < \infty$, and $||B||_{B(\ell_2)} = \sup_{||h||_2 \leq 1} ||V_{\mathcal{L}_B,h}||_\infty$.
v) There exists a continuous operator

$$\widetilde{T}_B : H_0^2([0,1]) \overline{\otimes} H_0^2([0,1]) \to C_s([0,1] \times [0,1]),$$

such that

$$\widetilde{T}_B \big|_{\mathcal{P}([0,1] \times [0,1])} = T_B .$$

Remark 2.9. *a) The equivalence i)-iv) above holds also in the more general case $B \in \mathcal{PM}(\ell_2)$.*

b) By the discussion above we have that:

$$||B||_{B(\ell_2)} = \sup_{||h||_2 \leq 1} \left(\sum_{k=1}^{\infty} \left| \int_0^1 \mathcal{L}_k^B(t)h(-t)dt \right|^2 \right)^{1/2}.$$

Finally, if $B \in M(\ell_2)$, $A \in B(\ell_2)$, $x = (x_n)_{n \geq 1} \in \ell_2(\mathbb{N})$, and $h \in H_0^2([0,1])$, with $\hat{h}(n) = x_n$, for all $n \geq 1$, then we have that

$$||(B*A)x||_2^2 = \sum_{k=1}^{\infty} \left| \int_0^1 \left(\mathcal{L}_k^B * \mathcal{L}_k^A \right)(t)h(-t)dt \right|^2.$$

Hence, we find that

$$||B||_{M(\ell_2)} = \sup_{||\mathcal{L}_A||_{H_0^2(\infty)} \leq 1} \sup_{||h||_2 \leq 1} \left(\sum_{k=1}^{\infty} \left| \int_0^1 \left(\mathcal{L}_k^B * \mathcal{L}_k^A \right)(t)h(-t)dt \right|^2 \right)^{1/2}.$$

2.2 Nonperiodical integral operators

In what follows we present another method to use the functions in the framework of matrix theory.

Let $B = (b_{ij})_{i,j \geq 1} \in B(\ell_2)$ and $x \in \ell_2(\mathbb{N}^*)$. Then

$$||Bx||_2^2 = \sum_{k=1}^{\infty} |b_{k1}x_1 + b_{k2}x_2 + \ldots|^2.$$

We define

$$\mathcal{P}(0,\infty) :=$$

$$\{f : (0,\infty) \to \mathbb{C} \text{ a measurable function } f\big|_{(k,k+1]} = ct \text{ a.e. } \forall k \in \mathbb{N}\},$$

and

$$\mathcal{L}^2(0,\infty) := \mathcal{P}(0,\infty) \cap L^2(0,\infty).$$

Then we have a one-to-one correspondence between the class of all sequences from $\ell_2(\mathbb{N}^*)$ and the space $\mathcal{L}^2(0,\infty)$, given by

$$x = (x_k)_{k \geq 1} \in \ell_2(\mathbb{N}^*) \leftrightarrow h \in \mathcal{L}^2(0,\infty),$$

with

$$h\big|_{(k,k+1]} = x_{k+1} \quad \forall k \in \mathbb{N}.$$

Definition 2.10. *Let* $B \in \mathcal{PM}(\ell_2)$. *To* B *it corresponds a unique function* $C^B(\cdot,\cdot)$ *given by*

  i) $C^B : (0,\infty) \times (0,\infty) \to \mathbb{C}$,

  ii) $C^B(\cdot,t) \in \mathcal{P}(0,\infty)$, *and* $C^B(y,\cdot) \in \mathcal{P}(0,\infty)\, \forall t, y > 0$,

  iii) $C^B(y,t) = b_{kj}$ *if* $y \in (k-1,k]$ *and* $t \in (j-1,j]$, *for all* $k, j \geq 1$,

where $B = (b_{kj})_{k \geq 1;\, j \geq 1}$.

Then, we have that

$$||Bx||_2^2 = \sum_{k=1}^{\infty} \left| \int_0^{\infty} C^B(k,t)h(t)dt \right|^2 = \int_0^{\infty} \left| \int_0^{\infty} C^B(y,t)h(t)dt \right|^2 dy.$$

Proposition 2.11. *Let* $B \in \mathcal{PM}(\ell_2)$ *and* $C^B(\cdot,\cdot)$ *be its associated function. The following assertions are equivalent:*

  i) $B \in B(\ell_2)$.

  ii) The operator

$$T_B : \mathcal{L}^2(0,\infty) \to \mathcal{L}^2(0,\infty),$$

given by

$$T_B(h)(y) = \int_0^{\infty} C^B(y,t)h(t)dt$$

is a continuous operator.

2.3   Some applications of integral operators in the classical theory of infinite matrices

In this section we use the previous results to give different proofs for some classical theorems of infinite matrix theory.

2.3.1   *The characterization of Toeplitz matrices*

We recall the definition of Toeplitz matrices.

Definition 2.12. *Let* $A \in \mathcal{PM}(\ell_2)$. *The matrix* A *is a Toeplitz matrix if there exists a sequence of complex numbers* $(a_n)_{n \in \mathbb{Z}}$ *such that*

$$A = \begin{pmatrix} a_0 & a_1 & a_2 & \cdots & a_n & \cdots \\ a_{-1} & a_0 & a_1 & \cdots & a_{n-1} & \cdots \\ \vdots & \vdots & \vdots & \vdots & \vdots & \cdots \\ a_{-n} & a_{-n+1} & a_{-n+2} & \cdots & a_0 & \cdots \\ \vdots & \vdots & \vdots & \vdots & \vdots & \cdots \end{pmatrix}. \tag{2.1}$$

To such a matrix it is possible to associate a unique pseudomeasure f_A given by

$$f_A(t) = \sum_{k=-\infty}^{+\infty} a_k e^{2\pi ikt},$$

equality being taken in the distribution's sense, and $t \in [0,1]$.

In what follows we find the necessary and sufficient conditions in order that a Toeplitz matrix A belong to $B(\ell_2)$, resp. to $M(\ell_2)$.

More specifically, we have:

Theorem 2.13. *Let A be a Toeplitz matrix like in (2.1). Then*

$$A \in B(\ell_2) \Leftrightarrow f_A \in L^\infty([0,1]).$$

Proof. By Proposition 2.7 we have that, if T_A is the matriceal operator associated to A, given by

$$T_A : \mathcal{P}([0,1]) \otimes \mathcal{P}([0,1]) \to C([0,1] \otimes [0,1]),$$

$$T_A(r \otimes h)(u,v) = \int_0^1 \int_0^1 \mathcal{L}_A(t,x) r(u-x) h(v-t) dx dt,$$

then the following assertions are equivalent:

1) $A \in B(\ell_2)$.

2) $||T_A(r \otimes h)||_\infty \le C||r||_2||h||_2$, where $C > 0$ is an absolute constant and $r, h \in H_0^2([0,1])$ are arbitrary functions.

Consequently, it is enough to prove that

$$2) \Leftrightarrow f_A \in L^\infty([0,1]).$$

Therefore, let us consider the pseudomeasures

$$\mathcal{L}_k^A(t) = \sum_{j=1}^\infty a_{-k+j} e^{2\pi ijt} = \mathcal{L}_k^{*\,A}(t) e^{2\pi ikt} \quad \forall\, k \ge 1.$$

The matriceal distribution associated to A is

$$\mathcal{L}_A(t,x) = \sum_{k=1}^\infty \mathcal{L}_k^A(t) \cdot e^{2\pi ikx}.$$

For $r, h \in \mathcal{P}([0,1])$ we have that

$$T_A(r \otimes h)(u,v) = \sum_{k=1}^\infty \int_0^1 \int_0^1 \mathcal{L}_k^{*\,A}(t) e^{2\pi ik(t+x)} r(u-x) h(v-t) dx dt$$

$$= \sum_{k=1}^{\infty} \left(\int_0^1 r(u-x)e^{2\pi i k x} dx \right) \left(\int_0^1 \mathcal{L}_k^{*A}(t)e^{2\pi i k t} h(v-t)dt \right)$$

$$= \sum_{k=1}^{\infty} \left(\int_0^1 r(x)e^{-2\pi i k x} dx \right) \left(\int_0^1 \mathcal{L}_k^{*A}(t)e^{2\pi i k t} h(v-t)dt \right) e^{2\pi i k u}.$$

$$\int_0^1 \mathcal{L}_k^{*A}(t)e^{2\pi i k t} h(v-t)dt = \int_0^1 f_A(t)e^{2\pi i k t} h(v-t)dt \quad \text{for all } k \geq 1.$$

Hence,

$$T_A(r \otimes h)(u,v) =$$

$$\int_0^1 f_A(t)h(v-t) \left(\sum_{k=1}^{\infty} \left(\int_0^1 r(x)e^{-2\pi i k x} dx \right) e^{2\pi i k(u+t)} \right) dt$$

$$= \int_0^1 f_A(t)h(v-t)r(u+t)dt.$$

In this way we have to prove the equivalence of the following two conditions:

i) $f_A \in L^{\infty}([0,1])$.

ii) There exists a constant $C > 0$ such that, for all $h, r \in \mathcal{P}([0,1])$, we have that

$$\left|\left| \int_0^1 f_A(t)h(\cdot - t)r(\cdot + t)dt \right|\right|_{\infty} \leq C||h||_2||r||_2.$$

$i) \Rightarrow ii)$ By Schwarz inequality and a trivial estimate we have that

$$\left| \int_0^1 f_A(t)h(v-t)r(u+t)dt \right| \leq ||f_A||_{\infty} \int_0^1 |h(v-t)r(u+t)|dt \leq$$

$$||f_A||_{\infty}||h||_2||r||_2, \text{ for fixed } u,v \in [0,1].$$

$ii) \Rightarrow i)$ Since

$$\sup_{||h||_2 \leq 1} \left| \int_0^1 f_A(t)h(-t)r(u+t)dt \right| = \left(\int_0^1 |f_A(t)r(u+t)|^2 dt \right)^{1/2},$$

by using ii) we have that:

$$\int_0^1 |f_A(t)r(t)|^2 dt \leq C \int_0^1 |r(t)|^2 dt \quad \forall r \in \mathcal{P}([0,1]).$$

Since

$$\sup_{||r||_2 \leq 1} \int_0^1 |f_A(t)r(t)|^2 dt = ||f_A||_{\infty}^2,$$

it follows that

$$||f_A||_{\infty}^2 \leq C,$$

i.e.

$$f_A \in L^{\infty}([0,1]). \qquad \square$$

We give a different proof of Bennett's Theorem (see [11]).

Theorem 2.14. *Let A be a Toeplitz matrix. Then*

$$A \in M(\ell_2) \Leftrightarrow f_A \in M([0,1]),$$

where this last space is the space of all bounded complex measures on $[0,1]$.

Proof. Let $B \in B(\ell_2)$ be fixed. It yields that

$$T_{A*B}(r \otimes h)(u,v) = \int_0^1 \int_0^1 \mathcal{L}_{A*B}(t,x)r(u-x)h(v-t)dxdt.$$

$$\mathcal{L}_{A*B}(t,x) = \sum_{k=1}^{\infty} \left(\mathcal{L}_k^A * \mathcal{L}_k^B\right)(t)e^{2\pi ikx} = \sum_{k=1}^{\infty} \left(\mathcal{L}_k^A * \mathcal{L}_k^B\right)^*(t)e^{2\pi ik(x+t)},$$

for $A \leftrightarrow (\mathcal{L}_k^A)_{k\geq1}$, and $B \leftrightarrow (\mathcal{L}_k^B)_{k\geq1}$.

Since

$$\left(\mathcal{L}_k^A * \mathcal{L}_k^B\right)^*(t) = \left(\mathcal{L}_k^{*A} * \mathcal{L}_k^{*B}\right)(t),$$

we have that

$$T_{A*B}(r \otimes h)(u,v)$$

$$= \sum_{k=1}^{\infty} \int_0^1 \int_0^1 \left(\mathcal{L}_k^{*A} * \mathcal{L}_k^{*B}\right)(t)e^{2\pi ik(x+t)}r(u-x)h(v-t)dxdt$$

$$= \sum_{k=1}^{\infty} \int_0^1 \int_0^1 \left(f_A * \mathcal{L}_k^{*B}\right)(t)e^{2\pi ik(x+t)}r(u-x)h(v-t)dxdt$$

$$= \int_0^1 f_A(s) \left(\int_0^1 \int_0^1 \underbrace{\sum_{k=1}^{\infty} \mathcal{L}_k^B(t-s)e^{2\pi ik(x+s)}}_{\mathcal{L}_B(t-s,x+s)} r(u-x)h(v-t)dxdt \right) ds$$

$$= \int_0^1 T_B(r \otimes h)(u+s,v-s)ds.$$

Thus, we conclude that

$$T_{A*B}(r \otimes h)(u,v) = \int_0^1 T_B(r \otimes h)(u+s,v-s)ds,$$

for a Toeplitz matrix A and arbitrary $B \in B(\ell_2)$.

Moreover, by using Theorem 2.13, we find that

$$T_B(r \otimes h)(u,v) = \int_0^1 f_B h(v-t) r(u+t) dt,$$

for a Toeplitz matrix $B \in B(\ell_2)$.

$\Leftarrow$ Suppose that $f_A \in M([0,1])$. By using Proposition 2.7 it is enough to show that:

$\forall B \in B(\ell_2), \exists C_B > 0$ such that $\forall r, h \in \mathcal{P}([0,1])$, it follows that

$$||T_{A*B}(r \otimes h)||_\infty \le C_B ||r||_2 ||h||_2. \tag{2.2}$$

But, for $B \in B(\ell_2)$,

$$T_B : \mathcal{P}([0,1]) \otimes \mathcal{P}([0,1]) \to C([0,1] \times [0,1])$$

can be extended continuously to an operator $\tilde{T}_B$.

Hence, we have that

$$T_{A*B}(r \otimes h)(u,v) = \int_0^1 f_A(s) T_B(r \otimes h)(u+s, v-s) ds,$$

and, since $f_A \in M([0,1])$, it follows that

$$|T_{A*B}(r \otimes h)(u,v)| \le ||f_A||_{M([0,1])} ||T_B(r \otimes h)||_\infty.$$

But

$$||T_B(r \otimes h)||_\infty \le C_B^1 ||r||_2 ||h||_2.$$

Consequently,

$$||T_{A*B}(r \otimes h)||_\infty \le C_B ||r||_2 ||h||_2,$$

where $C_B = C_B^1 ||f_A||_M$.

$\Rightarrow$ Let A, B be Toeplitz matrices with $A \in M(\ell_2)$, and $B \in B(\ell_2)$, $||B||_{B(\ell_2)} \le 1$.

Then, we have that

$$T_B(r \otimes h)(u,v) = \int_0^1 f_B(t) r(u+t) h(v-t) dt,$$

and, by using Theorem 2.13 and the relation (2.2), it follows that there exists $C > 0$ such that, for each $f_B \in L^\infty([0,1])$, with $||f_B||_\infty \le 1$, we have, $\forall r, h \in \mathcal{P}([0,1])$, that

$$\left| \int_0^1 f_A(s) \int_0^1 f_B(t-s) h(-t) r(t) dt ds \right| \le C ||h||_2 ||r||_2. \tag{2.3}$$

Next we recall two notions from classical Fourier analysis: the Cesaro kernel

$$K_n(t) = \sum_{j=-n}^{n} \left(1 - \frac{|j|}{n+1}\right) e^{2\pi i j t} = \frac{1}{n+1} \left[\frac{\sin(n+1)\pi t}{\sin \pi t}\right]^2 \geq 0,$$

and the Dirichlet kernel

$$D_n(t) = \sum_{j=-n}^{n} e^{2\pi i j t} = \frac{\sin \pi (2n+1)t}{\sin \pi t}.$$

We introduce in relation (2.3)

$$h(t) = \frac{1}{\sqrt{2n+1}} D_n(t) e^{2\pi i (n+1)t} = r(t).$$

Then $r, h \in \mathcal{P}([0,1])$, with $||h||_2 = ||r||_2 = 1$, and (2.3) becomes

$$\left|\int_0^1 f_A(s)\sigma_{2n+1}(f_B)(-s)ds\right| \leq C \quad \forall n \geq 1,$$

where

$$\sigma_n(f_B)(s) = (K_n * f_B)(s)$$

are the Cesaro sums of the order n associated to f_B.

Then, for all $n \geq 1$, and for all $f_B \in L^\infty([0,1])$, with $||f_B||_\infty \leq 1$, we have that

$$\left|\int_0^1 f_B(-s)\sigma_{2n+1}(f_A)(s)ds\right| \leq C, \tag{2.4}$$

and, hence, for all $n \geq 1$,

$$||\sigma_{2n+1}(f_A)||_1 = \sup_{||f_B||_\infty \leq 1} \left|\int_0^1 f_B(-s)\sigma_{2n+1}(f_A)(s)ds\right| \leq C.$$

Moreover, we define the sequence of functionals

$$S_n : C([0,1]) \to \mathbb{C},$$

by

$$S_n(g) = \int_0^1 \sigma_{2n+1}(f_A)(s)g(s)ds.$$

Then S_n are linear operators with $||S_n|| \leq C$ for all $n \geq 1$, and, by Alaoglu's Theorem, there exists a linear bounded operator

$$S : C([0,1]) \to \mathbb{C}$$

such that

$$S_n \to S \text{ weakly } .$$

Applying now Riesz Theorem, there exists $\mu \in M([0,1])$ such that

$$S(g) = \int_0^1 g d\mu.$$

It is clear that

$$\mu = f_A \in M([0,1]). \qquad \square$$

2.3.2   The characterization of Hankel matrices

Definition 2.15.   *Let $A \in \mathcal{PM}(\ell_2)$. Then A is called a Hankel matrix if there exists the complex sequence $(a_n)_{n \in \mathbb{N}^*}$ such that*

$$A = \begin{pmatrix} a_1 & a_2 & a_3 & \nearrow & a_n \\ a_2 & a_3 & \nearrow & a_n & \nearrow \\ a_3 & \nearrow & a_n & \nearrow & \\ & \nearrow & a_n & \nearrow & \\ a_n & \nearrow & & & \\ & \nearrow & & & \end{pmatrix}.$$

To such a matrix we associate a unique pseudomeasure g_A given by

$$g_A(t) = \sum_{k=1}^{\infty} a_k e^{2\pi i k t} \text{ where } t \in [0,1].$$

Like in the case of Toeplitz matrices we study whenever the matrix A belongs to the spaces $B(\ell_2)$, or $M(\ell_2)$.

More specifically, we have that

Theorem 2.16. *Let A be a Hankel matrix. Then*

$$A \in B(\ell_2) \Leftrightarrow g_A \in BMO.$$

Proof.   Let T_A be the matriceal operator associated to A

$$T_A(r \otimes h)(u,v) = \int_0^1 \int_0^1 \mathcal{L}_A(t,x) r(u-x) h(v-t) dx dt$$

$$= \sum_{k=1}^{\infty} \left(\int_0^1 \mathcal{L}_k^A(t) h(v-t) dt \right) \left(\int_0^1 r(u-x) e^{2\pi i k x} dx \right).$$

Since

$$\int_0^1 \mathcal{L}_k^A(t) h(v-t) dt = \int_0^1 g_A(t) e^{-2\pi i (k-1) t} h(v-t) dt,$$

we have that

$$T_A(r \otimes h)(u,v) =$$

$$\int_0^1 g_A(t) e^{2\pi i t} h(v-t) \left(\sum_{k=1}^{\infty} \left(\int_0^1 r(x) e^{-2\pi i k x} dx \right) e^{2\pi i k (u-t)} \right) dt.$$

Thus,

$$T_A(r \otimes h)(u,v) = \int_0^1 g_A(t)h(v-t)e^{2\pi it}r(u-t)dt, \qquad (2.5)$$

or, denoting by $h_v(t) = h(v+t)$, and by $r_u(t) = r(u+t)$,

$$T_A(r \otimes h)(u,v) = \int_0^1 g_A(t)e^{2\pi it}(h_v r_u)(-t)dt.$$

Next we prove the implication $\Rightarrow$.

If $A \in B(\ell_2)$, we have that, $r, h \in \mathcal{P}([0,1])$,

$$|T_A(r \otimes h)(0,0)| = \left| \int_0^1 g_A(t)e^{2\pi it}(hr)(-t)dt \right| \leq C||r||_2||h||_2.$$

In view of the theory of Hardy spaces, this last fact is equivalent to the following inequality:

$$\sup_{f \in H_0^1, ||f||_1 \leq 1} \left| \int_0^1 g_A(t)e^{2\pi it}f(-t)dt \right| \leq C,$$

i.e., to that

$$||g_A||_{BMO} \leq C.$$

For the converse implication let $g_A \in BMO$. We have that

$$|T_A(r \otimes h)(u,v)| = \left| \int_0^1 g_A(t)e^{2\pi it}(h_v r_u)(-t)dt \right| \leq ||g_A||_{BMO}||h_v r_u||_{H^1}$$

$$\leq ||g_A||_{BMO}||h||_2||r||_2,$$

i.e.,

$$||T_A|| \leq ||g_A||_{BMO}. \qquad \square$$

Theorem 2.17. *Let A be a Hankel matrix. We have:*

i) $A \in M(\ell_2)$ implies that $g_A \in M(H_0^1, H_0^1)$, where this last space is the space of all Fourier multipliers of H_0^1.

ii) $g_A \in M([0,1])$ implies that $A \in M(\ell_2)$.

Proof. Let $B \in B(\ell_2)$. Then

$$T_{A*B}(r \otimes h)(u,v) = \sum_{k=1}^\infty \int_0^1 \int_0^1 \left(\mathcal{L}_k^A * \mathcal{L}_k^B \right)(t)h(v-t)r(u-x)e^{2\pi ikx}dtdx$$

$$= \int_0^1 g_A(s)e^{2\pi is} \left(\int_0^1 \int_0^1 \mathcal{L}_k^B(t-s)e^{2\pi ik(x-s)}h(v-t)r(u-x)dtdx \right) ds.$$

Thus,

$$T_{A*B}(r \otimes h)(u, v) = \int_0^1 g_A(s)e^{2\pi is}T_B(r \otimes h)(u - s, v - s)ds. \qquad (2.6)$$

i) We assume that $A \in M(\ell_2)$. Then there exists $C > 0$ such that, for each Hankel matrix B, $B \in B(\ell_2)$, with $||B||_{B(\ell_2)} \leq 1$, we have that, for $r, h \in \mathcal{P}([0,1])$,

$$||T_{A*B}(r \otimes h)||_\infty \leq C||r||_2||h||_2. \qquad (2.7)$$

Therefore, by using the relations (2.5), (2.6), and (2.7), we have that

$$\left| \int_0^1 g_A(s)e^{2\pi is} \left(\int_0^1 g_B(t)e^{2\pi it}(r_{u-s}h_{v-s})(-t)dt \right) ds \right| \leq C||r||_2||h||_2.$$

We take $u = v = 0$, and $s + t = y$ in the relation above, obtaining that

$$\left| \int_0^1 g_A(s) \left(\int_0^1 g_B(y - s)e^{2\pi iy}(rh)(-y)dy \right) ds \right| \leq C||r||_2||h||_2,$$

or, equivalently,

$$\sup_{g_B \in BMO; ||g_B||_{BMO} \leq 1} \sup_{f \in H_0^1; ||f||_1 \leq 1} \left| \int_0^1 g_B(s) \left(\int_0^1 g_A(y - s)f(-y)dy \right) ds \right| \leq C.$$

Thus, the operator

$$S_A : H_0^1 \to H_0^1,$$

given by

$$S_A(f)(s) = \int_0^1 g_A(s - y)f(y)dy,$$

is a bounded operator if and only if

$$g_A \in M(H_0^1, H_0^1).$$

ii) According to relation (2.7), for every $B \in B(\ell_2)$, and for $g_A \in M([0,1])$, we have that

$$||T_{A*B}(r \otimes h)||_\infty \leq ||g_A||_M ||T_B(r \otimes h)||_\infty.$$

Since $B \in B(\ell_2)$, it follows that

$$||T_B(r \otimes h)||_\infty \leq C_B||r||_2||h||_2,$$

therefore,

$$||T_{A*B}(r \otimes h)||_\infty \leq ||g_A||_M \tilde{C}_B||r||_2||h||_2.$$

Hence,

$$A * B \in B(\ell_2) \quad \forall B \in B(\ell_2),$$

that is

$$A \in M(\ell_2). \qquad \square$$

Remark 2.18. *An equivalent condition to the statement that a Hankel matrix $A \in M(\ell_2)$ was given by G. Pisier in [81]:*

$$A \in M(\ell_2) \Leftrightarrow g_A \in M(H^1(S^1), H^1(S^1)),$$

where $H^1(S^1)$ is the operator trace class-valued Hardy space.

2.3.3 *The main triangle projection*

In what follows we present a new proof of a result in [49], namely:
What is the growth rate with respect to n of the expression

$$\sup_{\|A\|_{B(\ell_2)}\leq 1} \|P_n(A)\|_{B(\ell_2)},$$

where

$$P_n(A) = (b_{ij})\ i,j \geq 1,\ for\ b_{ij} = \begin{cases} a_{ij} & if\ i+j \leq n+1 \\ 0 & otherwise, \end{cases}$$

with $A = (a_{ij})_{i,j\geq 1}$.

Theorem 2.19. *Let P_n the triangle projection of the order n. Then we have that*

$$\sup_{\|A\|_{B(\ell_2)}\leq 1} \|P_n(A)\|_{B(\ell_2)} = \mathcal{O}(\log n) \quad for\ n \to \infty.$$

Moreover, there exists $C > 0$ such that

$$\sup_{\|A\|_{B(\ell_2)}\leq 1} \|P_n(A)\|_{B(\ell_2)} \geq C \log n \quad \forall n \in \mathbb{N}^*.$$

Proof. Let $A \in B(\ell_2)$ and $x \in \ell_2(\mathbb{N})$. Then, by using the correspondences $A \leftrightarrow (\mathcal{L}_k^A)_{k\geq 1}$, and $x = (x_j)_{j\geq 1} \leftrightarrow h(t) = \sum_{j=1}^{\infty} x_j e^{2\pi ijt}$, as in Section 2.1, we have the following equality:

$$\|P_n(A)x\|_2^2 = \sum_{k=1}^{n} \left| \int_0^1 \left(\mathcal{L}_k^A * D_{n+1-k}\right)(t)h(-t)dt \right|^2,$$

where

$$D_n(t) = \sum_{k=-n}^{n} e^{2\pi ikt}$$

is the Dirichlet's kernel.
Therefore,

$$\|P_n(A)x\|_2^2 = \sum_{k=1}^{n} \left| \int_0^1 D_{n+1-k}(s) \left(\int_0^1 \mathcal{L}_k^A(t-s)h(-t)dt \right) ds \right|^2$$

$$\leq \sum_{k=1}^{n} \|D_{n+1-k}\|_1 \int_0^1 |D_{n+1-k}(s)| \left|\left(\mathcal{L}_k^A * h\right)(-s)\right|^2 ds$$

$$\leq \sup_{1\leq k\leq n} \|D_{n+1-k}\|_1 \int_0^1 \left(\sup_{1\leq k\leq n} |D_{n+1-k}(s)| \right) \left(\sum_{k=1}^{n} \left|\left(\mathcal{L}_k^A * h\right)(-s)\right|^2 \right) ds$$

$$\leq \sup_{1 \leq k \leq n} ||D_{n+1-k}||_1 \left|\left| \sup_{1 \leq k \leq n} |D_{n+1-k}(s)| \right|\right|_1 ||V_{\mathcal{L}_A,h}||_\infty^2 .$$

Since

$$||D_n||_1 = \mathcal{O}(\log n),$$

for $n \to \infty$, it remains to prove that

$$\int_0^1 \sup_{1 \leq k \leq n} |D_k(s)| \, ds = \mathcal{O}(\log n) \Leftrightarrow \int_0^1 \sup_{1 \leq k \leq n} \frac{|\sin ks|}{s} ds = \mathcal{O}(\log n).$$

In order to prove this we take $g_s : [1, u] \to \mathbb{R}_+$, given by

$$g_s(x) = \frac{|\sin xs|}{s},$$

where $s \in (0, 1]$.

There are two distinct cases:

1) $s > \frac{1}{2n}$, which implies that

$$g_s(x) < \frac{1}{s} \quad \forall x \in [1, n],$$

and

2) $s \leq \frac{1}{2n}$ which implies $0 < xs \leq \frac{n}{2n} = \frac{1}{2}$, so that

$$g_s(x) \leq \frac{\sin ns}{s} \quad \forall x \in [1, n].$$

Therefore,

$$\int_0^1 \sup_{1 \leq k \leq n} \frac{|\sin ks|}{s} ds \leq \int_0^{\frac{1}{2n}} \frac{\sin ns}{s} ds + \int_{\frac{1}{2n}}^1 \frac{1}{s} ds$$

$$= \int_0^{\frac{1}{2}} \frac{\sin s}{s} ds + \log 2n = \mathcal{O}(\log n),$$

for n sufficiently large, which, in turn, implies that

$$||P_n(A)||_{B(\ell_2)} \leq C \log n \sup_{||h||_2 \leq 1} ||V_{\mathcal{L}_A,h}||_\infty.$$

According to Remark 2.2 we have that

$$||P_n(A)||_{B(\ell_2)} \leq C \log n ||A||_{B(\ell_2)}.$$

In order to prove that there exists $B > 0$, such that

$$\sup_{||A||_{B(\ell_2)} \leq 1} ||P_n(A)||_{B(\ell_2)} \geq B \log n,$$

we observe that

$$\sup_{||A||_{B(\ell_2)} \leq 1} ||P_n(A)||_{B(\ell_2)} = \sup_{||A||_{B(\ell_2)} \leq 1} ||T_n(A)||_{B(\ell_2)},$$

where $A = (a_{ij})_{i,j \geq 1}$, and

$$T_n(A) = \begin{matrix} \\ \\ \\ \\ n \\ n+1 \\ \\ \end{matrix} \begin{pmatrix} a_{11} & 0 & 0 & \ldots & \overset{n}{0} & \overset{n+1}{0} & \ldots \\ a_{21} & a_{22} & 0 & \ldots & 0 & 0 & \ldots \\ \vdots & \vdots & \vdots & \vdots & \vdots & \vdots & \ldots \\ a_{n1} & a_{n2} & a_{n3} & \ldots & a_{nn} & 0 & \ldots \\ 0 & 0 & 0 & 0 & 0 & 0 & \ldots \\ \vdots & \vdots & \vdots & \vdots & \vdots & \vdots & \ldots \end{pmatrix}.$$

We consider next Hilbert's matrix given by

$$H = \begin{pmatrix} 0 & 1 & \frac{1}{2} & \frac{1}{3} & \ldots \\ -1 & 0 & 1 & \frac{1}{2} & \ldots \\ -\frac{1}{2} & -1 & 0 & 1 & \ldots \\ \vdots & \ddots & \ddots & \ddots & \ddots \end{pmatrix},$$

and $x_n = (\overset{1}{1}, \overset{2}{1}, \ldots, \overset{n}{1}, 0, \ldots)$.

Then

$$||H||_{B(\ell_2)} < \infty,$$

and

$$||T_n(H)x_n||_2^2 = 1^2 + \left(1 + \frac{1}{2}\right)^2 + \cdots + \left(1 + \frac{1}{2} + \cdots + \frac{1}{n-1}\right)^2$$

$$\geq C \left(\log^2 2 + \log^2 3 + \cdots + \log^2 n\right) = C \int_2^n \log^2 x \, dx \sim C n \log^2 n.$$

Therefore

$$||T_n(H)||_{B(\ell_2)}^2 \geq \frac{||T_n(H)x_n||_2^2}{||x_n||_2^2} \geq C \frac{n \log^2 n}{n},$$

that is

$$\sup_{||A||_{B(\ell_2)} \leq 1} ||T_n(A)||_{B(\ell_2)} \geq ||H||_{B(\ell_2)}^{-1} ||T_n(H)||_{B(\ell_2)} \geq C \log n.$$

$\square$

2.3.4 $B(\ell_2)$ is a Banach algebra under the Schur product

Next we give a different proof of an old result of I. Schur (see also [11]).

Theorem 2.20. $B(\ell_2, *)$ *is a Banach algebra.*

Proof.   Let $A, B \in B(\ell_2)$, and $(\mathcal{L}_k^A)_{k \geq 1}$, $(\mathcal{L}_k^B)_{k \geq 1}$, be the sequences of functions associated to A, and to B, respectively.

We have that

$$||(B * A)x||_2^2 = \sum_{k=1}^{\infty} \left| \int_0^1 \left(\mathcal{L}_k^A * \mathcal{L}_k^B \right)(t)h(-t)dt \right|^2$$

$$= \sum_{k=1}^{\infty} \left| \int_0^1 \mathcal{L}_k^A(s) \left(\int_0^1 \mathcal{L}_k^B(t - s)h(-t)dt \right) ds \right|^2$$

$$\leq \sum_{k=1}^{\infty} \left(\int_0^1 \left| \mathcal{L}_k^A(s) \right|^2 ds \right) \left(\int_0^1 \left| \left(\mathcal{L}_k^B * h \right)(-s) \right|^2 ds \right),$$

where $x = (x_j)_{j \geq 1} \in \ell_2(\mathbb{N})$, and

$$h(t) = \sum_{j=1}^{\infty} x_j e^{2\pi i j t} \in H_0^2([0, 1]).$$

Therefore,

$$||(B * A)||_2 \leq \sup_{k \geq 1} \left(\int_0^1 \left| \mathcal{L}_k^A(s) \right|^2 ds \right)^{1/2} \left(\int_0^1 V_{\mathcal{L}_B, h}^2(s) ds \right)^{1/2}$$

$$\leq ||A||_{B(\ell_2)} ||V_{\mathcal{L}_B, h}||_2.$$

Finally,

$$||B * A||_{B(\ell_2)} \leq ||A||_{B(\ell_2)} \sup_{||h||_2 \leq 1} ||V_{\mathcal{L}_B, h}||_2 \leq ||A||_{B(\ell_2)} ||B||_{B(\ell_2)}. \qquad \square$$

Notes

The main idea of this chapter is the interpretation of an infinite matrix as a sequence of functions (or, more generally, distributions). We take in this way the advantage of a rich class of notions and techniques, which are usual in function theory. For instance, in Section 2.1 we define the notions of a square function, respectively of a matriceal operator associated to a given matrix A.

The main result of Section 2.1 is of course Theorem 2.8.

In Section 2.2 the central result is Proposition 2.11, which is the non-periodical analogue of Theorem 2.8.

Section 2.3 is dedicated to applications of results from Section 2.1. In this way we give different proofs of some classical results from infinite matrix theory: Theorem 2.13 is of course well-known (see [94]). Theorem 2.14 is known as Bennett's Theorem (see [11]). Theorem 2.16, known as Nehari's Theorem (see [65]), is here presented with a new proof. Finally, we mention Theorem 2.19, first proved in [49]. Theorem 2.20 was discovered apparently by I. Schur (see [11] for a more general result).

Chapter 3

Matrix versions of spaces of periodical functions

An interesting problem concerning infinite matrices is the following:

Let $A \in B(\ell_2)$. When is the matrix A approximable by matrices of finitely band type in the operator norm $|| \cdot ||_{B(\ell_2)}$?

In what follows we deal with this problem in some spaces of infinite matrices which can be regarded as extensions of classical Banach spaces of functions $C(\mathbb{T})$ and $L^1(\mathbb{T})$. These Banach spaces together with the spaces $B(\ell_2)$ and $M(\ell_2)$ are of interest in order to develop some results extending known theorems of classical harmonic analysis in the framework of matrices.

One main aim of the present chapter is to extend in the framework of matrices Fejer's theory for Fourier series. (See for instance [96].)

As it was stated in the Introduction we have a similarity between the expansion in the Fourier series $f = \sum_k a_k e^{ikx}$ of a periodical function f on the torus $\mathbb{T}$ and the decomposition $A = \sum_{k \in \mathbb{Z}} A_k$, where A_k is the kth diagonal of A for $k \in \mathbb{Z}$.

Moreover, there is a similarity between the convolution product $f * g$ of two periodical functions and the Schur product of two matrices A and B, $C = A * B$.

First we mention the following results obtained by Fejer, which have been guiding for our investigation:

(A) A function $f(\theta) = \sum_{k \in \mathbb{Z}} a_k e^{ik\theta}$ is continuous on $\mathbb{T}$ (that is $f \in C(\mathbb{T})$) if and only if the Cesaro sums

$$\sigma_n(f) = \sum_{k=-n}^{k} a_k \left(1 - \frac{|k|}{n+1} \right) e^{ik\theta}$$

converge uniformly on $\mathbb{T}$ to f.

(B) A function $f(\theta) = \sum_{k \in \mathbb{Z}} m_k e^{ik\theta} \in L^1(\mathbb{T})$ if and only if
$$||\sigma_n(f) - f||_{L^1(\mathbb{T})} \to 0 \quad as \quad n \to \infty.$$

3.1   Preliminaries

In view of Fejer's result (A) it is natural to give the following definition:

Definition 3.1. *Let $A \in B(\ell_2)$. We denote by $\sigma_n(A)$ the Cesaro sum associated to $S_n(A) := \sum_{k=-n}^{n} A_k$, that is $\sigma_n(A) = \sum_{k=-n}^{n} A_k \left(1 - \frac{|k|}{n+1}\right)$. Then we say that A is a* continuous matrix *if*

$$\lim_{n \to \infty} ||\sigma_n(A) - A||_{B(\ell_2)} = 0.$$

Let us denote by $C(\ell_2)$ the vector space of all *continuous matrices* and consider on it the usual operator norm.

Now recall that the space of all Schur multipliers $M(\ell_2)$ is a commutative unital Banach algebra with respect to Schur product.

Moreover we have Bennett's theorem (see for instance Theorem 2.14):

The Toeplitz matrix $M = (c_{j-k})_{j,k}$, where $(c_n)_{n \in \mathbb{Z}}$ is a sequence of complex numbers, is a Schur multiplier if, and only if, there exists a (bounded, complex, Borel) measure μ on (the circle group) $\mathbb{T}$ with

$$\widehat{\mu}(n) = c_n \quad \text{for } n = 0, \pm 1, \pm 2, \ldots.$$

Moreover, we then have

$$||M||_{M(\ell_2)} = ||\mu||.$$

We also mention the following well-known fact (see Theorem 2.13):

The Toeplitz matrix M represents a linear and bounded operator on ℓ_2 if and only if there exists a function $f \in L^{\infty}(\mathbb{T})$ with Fourier coefficients $\widehat{f}(n) = m_n$ for all $n \in \mathbb{Z}$. Moreover, we have

$$||M||_{B(\ell_2)} = ||f||_{L^{\infty}(\mathbb{T})}.$$

3.2   Some properties of the space $C(\ell_2)$

First of all let us observe the following fact:

Remark 3.2. *By Fejer's theorem (A) we have that a Toeplitz matrix $T = (t_k)_{k \in \mathbb{Z}} \in C(\ell_2)$ if and only if $f_T(\theta) \overset{def}{=} \sum_{k \in \mathbb{Z}} t_k e^{ik\theta} \in C(\mathbb{T})$, and in this way we can see that the notion of a continuous matrix may be regarded as an analogue of that of a continuous function.*

Now let C_∞ denote the space of all matrices defining compact operators.

Proposition 3.3. $C(\ell_2)$ *is a proper closed ideal of* $B(\ell_2)$ *with respect to Schur multiplication which, in its turn, contains* C_∞ *properly.*

Proof. We have:

$$||\sigma_n(A)||_{B(\ell_2)} = \left\| \sum_{k=-n}^{n} A_k \left(1 - \frac{|k|}{n+1}\right) \right\|_{B(\ell_2)} \leq ||M_n||_{M(\ell_2)} ||A||_{B(\ell_2)},$$

where M_n is the n-band type Toeplitz matrix with the entries

$$m_{ij} = \begin{cases} 1 - \dfrac{|j-i|}{n+1} & \text{if } |j-i| \leq n, \\ 0 & \text{otherwise.} \end{cases}$$

Hence $C(\ell_2)$ is a closed subspace of $B(\ell_2)$.

Now we observe that for $A, B \in C(\ell_2)$, $\sigma_n(A * B) = \sigma_n(A) * B$ and then we have that, for $A \in C(\ell_2)$, $B \in B(\ell_2)$

$$||A * B - \sigma_n(A * B)||_{B(\ell_2)} = ||[A - \sigma_n(A)] * B||_{B(\ell_2)}$$

$$\leq ||A - \sigma_n(A)||_{B(\ell_2)} \cdot ||B||_{M(\ell_2)} \leq ||B||_{B(\ell_2)} \cdot ||A - \sigma_n(A)||_{B(\ell_2)}.$$

Here we have used the simple fact that

$$||B||_{M(\ell_2)} = ||B * E||_{M(\ell_2)} \leq ||B||_{B(\ell_2)} \cdot ||E||_{M(\ell_2)} = ||B||_{B(\ell_2)},$$

$E = (E_{ij})$ where $E_{ij} = 1$ for all $i, j \in \mathbb{N}$, and $||E||_{M(\ell_2)} = 1$.

Hence, $C(\ell_2)$ is a closed ideal of $B(\ell_2)$ with respect to Schur multiplication.

Next we note that $C(\ell_2)$ is a proper ideal of $B(\ell_2)$.

Denoting by e_{ij} the matrix whose single non-zero entry is 1 on the ith row and on the jth column, we consider the matrix $A = \sum_{k \in \mathbb{N}} A_k$, where $A_k = e_{k+1, 2k+1}$, $k \geq 0$, which belongs to $B(\ell_2)$, since $(AA^*)^{1/2} = I$ (I is the identity matrix). Moreover,

$$||\sigma_n(A) - A||_{B(\ell_2)} = \left\| \sum_{k>n} A_k + \frac{1}{n+1} \sum_{k=0}^{n} k A_k \right\|_{B(\ell_2)} = \left(\max_{k \leq n} \frac{k}{n+1} \right) \vee 1 = 1$$

for all n and, thus, $A \notin C(\ell_2)$.

Now let $A \in C_\infty$. Denoting by

$$P_n(A)(i,j) = \begin{cases} a_{ij} & i, j \leq n \\ 0 & \text{otherwise,} \end{cases}$$

we have that

$$||P_n(A) - A||_{B(\ell_2)} \to 0, \text{ as } n \to \infty.$$

But, by Bennett's theorem, we have that for $k > n$:

$$||P_n(A) - \sigma_k(P_n(A))||_{B(\ell_2)} = \left\| \sum_{\ell=-n}^{n} (P_n(A))_\ell \frac{|\ell|}{k+1} \right\|_{B(\ell_2)}$$

$$\leq \left\| \sum_{\ell=-n}^{n} \frac{|\ell|}{k+1} e^{i\ell\theta} \right\|_{M(\mathbb{T})} \cdot ||P_n(A)||_{B(\ell_2)} \to 0 \text{ as } k \to \infty.$$

Hence, $P_n(A) \in C(\ell_2)$ for all $n \in \mathbb{N}$ and, consequently, $C_\infty \subset C(\ell_2)$.

Since it is easy to see and well-known that a Toeplitz matrix does not represent a compact operator, by the previous remark it follows that C_∞ is a proper subspace of $C(\ell_2)$. The proof is complete. □

3.3 Another characterization of the space $C(\ell_2)$ and related results

We will give another characterization of the space $C(\ell_2)$ by using continuous vector-valued functions but first we note the following simple fact:

Remark 3.4. *Consider the function* $f_A : \mathbb{T} \to B(\ell_2)$ *given by* $f_A(t) = A * \left(e^{i(j-k)t} \right)_{j,k\geq 0}$. *Then*

$$||f_A(\theta)||_{B(\ell_2)} = ||A||_{B(\ell_2)} \text{ for all } \theta \in \mathbb{T}.$$

Indeed, by Bennett's multiplier theorem, we have that

$$||f_A(\theta)||_{B(\ell_2)} \leq ||A||_{B(\ell_2)}||\delta_{-\theta}|| = ||A||_{B(\ell_2)},$$

where $\delta_\theta \in M(\mathbb{T})$ denote the Dirac point mass at $\theta \in \mathbb{T}$. Similarly, since $A = f_A(\theta) * \left(e^{i(j-k)t} \right)_{j,k\geq 0}$, we have that $||A||_{B(\ell_2)} \leq ||f_A(\theta)||_{B(\ell_2)}$.

An easy consequence of this remark is that f_A is continuous on $\mathbb{T}$ if and only if it is continuous at one single point.

Now we ask ourselves how the matrix A should be in order that the function f_A shall be continous.

The answer to this question is as follows:

Theorem 3.5. *Let A be an infinite matrix. Then f_A is a $B(\ell_2)$-valued continuous function if and only if $A \in C(\ell_2)$, with equality of the corresponding norms.*

Proof. By Remark 3.4 it follows that

$$||\sigma_n(f_A) - f_A||_{C(\mathbb{T},B(\ell_2))} = ||\sigma_n(A) - A||_{B(\ell_2)}.$$

Now reasoning as in the proof of Fejér's result (A) (see for instance [41]) we get that for a continuous function $f_A : \mathbb{T} \to B(\ell_2)$ it follows that

$$||\sigma_n(A) - A||_{B(\ell_2)} \to 0,$$

as $n \to \infty$. Thus $A \in C(\ell_2)$.

The converse implication follows easily from Remark 3.4. $\square$

Now we shall study the following question: *What can we say about subspaces of $M(\ell_2)$ in connection with the multiplier property?*

The following theorem gives a justification of introducing $C(\ell_2)$ and also a partial answer to the above question. It is the matriceal analogue of Theorem 11.10, Chap. IV-[96].

Theorem 3.6. *The Toeplitz matrix $M = (m_k)_{k\in\mathbb{Z}}$ is a Schur multiplier from $B(\ell_2)$ into $C(\ell_2)$ iff*

$$\sum_{k\in\mathbb{Z}} m_k e^{ik\theta} \in L^1(\mathbb{T}).$$

Proof. If we identify $f \in L^\infty(\mathbb{T})$ with its corresponding Toeplitz operator $T_f = \left(\widehat{f}(j-k)\right)_{j,k\geq 0}$ in $B(\ell_2)$, then it is straightforward to see that a Schur multiplier $M = (m_{j-k})_{j,k\geq 0}$ mapping $B(\ell_2)$ into $C(\ell_2)$ induces a Fourier multiplier sequence $m = \{m_n\}_{n=-\infty}^\infty$ mapping $L^\infty(\mathbb{T})$ into $C(\mathbb{T})$, which is known to correspond, in the manner indicated in the statement above, to a function from $L^1(\mathbb{T})$ (see [96]). The converse follows also by the same lines.

$\square$

Guided by [11] we propose for matrices a similar notion to that of Lebesgue integrable functions.

Definition 3.7. *We say that an infinite matrix A is an integrable matrix if $\sigma_n(A) \to A$ as $n \to \infty$ in the norm of $M(\ell_2)$. The space of all such matrices, endowed with the norm induced by $M(\ell_2)$, is denoted by $L^1(\ell_2)$.*

Of course $L^1(\ell_2)$ is a Banach space.

Remark 3.8. *If $A \in L^1(\ell_2)$, then it follows that $A * B \in C(\ell_2)$ for all $B \in B(\ell_2)$.*

Indeed, for $B \in B(\ell_2)$, in view of the fact that

$$||\sigma_n(A * B) - A * B||_{B(\ell_2)} \le ||\sigma_n(A) - A||_{M(\ell_2)} \cdot ||B||_{B(\ell_2)} \to 0,$$

as $n \to \infty$, we find that $A * B \in C(\ell_2)$.

Now it is clear that $L^1(\ell_2)$ is a closed ideal of $M(\ell_2)$ with respect to the Schur product.

We have the following analogue of the Riemann-Lebesgue Lemma:

Lemma 3.9. *Let $M \in L^1(\ell_2)$. Then*

$$\lim_{|k| \to \infty} ||M_k||_{L^1(\ell_2)} = 0.$$

Proof. It is clear from Definition 3.7 that, for any $\epsilon > 0$, there is a number $n(\epsilon)$ such that, for $|k| \ge n(\epsilon)$, it follows that $||M_k||_{L^1(\ell_2)} \le \epsilon$ and the proof is complete. $\square$

Remark 3.10. *It is easy to see that for a diagonal matrix A_k, $k \in \mathbb{Z}$, we have that*

$$||A_k||_{B(\ell_2)} = ||A_k||_{M(\ell_2)}.$$

Thus, in Lemma 3.9 we can take $||M_k||_{B(\ell_2)}$ instead of $||M_k||_{L^1(\ell_2)}$.

In view of Theorem 3.5 and Remark 3.8 it is natural to ask: *If A is a Schur multiplier which maps $B(\ell_2)$ into $C(\ell_2)$ does it follow that $A \in L^1(\ell_2)$?*

The answer to the above question is negative. In fact, we have that

Example 3.11. *Let A be the following matrix:*

$$A = \begin{pmatrix} 1 & 1 & 1 & \cdots \\ 0 & 0 & 0 & \cdots \\ \vdots & \vdots & \vdots & \ddots \end{pmatrix}.$$

The matrix A is a Schur multiplier which maps $B(\ell_2)$ into $C(\ell_2)$ but it does not belong to $L^1(\ell_2)$.

In fact, A is a Schur multiplier with the property that $A * B \in C(\ell_2)$ for all $B \in B(\ell_2)$ since the matrix $A * B$ has rank 1 and therefore represents a compact operator and consequently it belongs to $C(\ell_2)$. Moreover, A does not belong to $L^1(\ell_2)$ by Lemma 3.9.

Therefore the Banach space $(B(\ell_2), C(\ell_2))$ of all multipliers from $B(\ell_2)$ into $C(\ell_2)$ is different from both $M(\ell_2)$ and $L^1(\ell_2)$. Thus, it seems that this space deserves to be studied in more detail.

On the other hand the space $(C(\ell_2), C(\ell_2))$ of all infinite matrices A such that $A * B \in C(\ell_2)$ for all $B \in C(\ell_2)$ can be described easily. More precisely we have:

Theorem 3.12. $(C(\ell_2), C(\ell_2))$ *is exactly the space* $M(\ell_2)$ *of all Schur multipliers.*

Proof. Since $\sigma_n(A * M) = \sigma_n(A) * M$ it follows easily that $M \in (C(\ell_2), C(\ell_2))$ if $M \in M(\ell_2)$ and $A \in C(\ell_2)$.

Conversely, assuming that $M \in (C(\ell_2), C(\ell_2))$, we have for $A \in B(\ell_2)$ that $||M * \sigma_n(A)||_{B(\ell_2)} \leq C||\sigma_n(A)||_{B(\ell_2)}$.

Moreover, $\sigma_n(A) \to A$ in the weak topology of operators in $B(\ell_2)$, that is $< \sigma_n(A)x, y > \to < Ax, y >$ for all $x, y \in \ell_2$, where $< \cdot, \cdot >$ is the scalar product in ℓ_2 (use ℓ_2-sequences with a finite number of nonzero components and a standard approximation argument). This yields that $||M * A||_{B(\ell_2)} \leq C||A||_{B(\ell_2)}$, that is M is a Schur multiplier. The proof is complete. $\qquad\square$

Next we give a characterization of an integrable matrix in the spirit of Theorem 3.5:

Theorem 3.13. *Let* $A \in M(\ell_2)$ *and* $f_A(\theta) = A * (e^{i(j-k)})_{j,k \geq 0}$ *for* $\theta \in \mathbb{T}$. *Then* f_A *is a pointwise well-defined function* $f_A : \mathbb{T} \to M(\ell_2)$ *such that* $||f_A(\theta)||_{M(\ell_2)} = ||A||_{M(\ell_2)}$ *for all* $\theta \in \mathbb{T}$. *Furthermore,* $f_A \in C(\mathbb{T}, M(\ell_2))$, *that is* f_A *is continuous, if and only if* $A \in L^1(\ell_2)$.

Proof. For $\mu \in M(\mathbb{T})$ let us introduce the notation T_μ for the Toeplitz matrix with symbol μ, that is

$$T_\mu = (\widehat{\mu}(j - k))_{j,k \geq 0}, \text{ where } \widehat{\mu}(n) = \int e^{-int} d\mu(t)$$

is the nth Fourier coefficient of μ. Note that $f_A(\theta) = A * T_{\delta_0}$, where $\delta_0 \in M(\mathbb{T})$ denotes the unit point mass at $\theta \in \mathbb{T}$. By Bennett's multiplier theorem (see [11]) we obtain that

$$||f_A(\theta)||_{M(\ell_2)} \leq ||A||_{M(\ell_2)}.$$

Similarly, since $A = f_A(\theta) * T_{\delta_0}$, we have that $||A||_{M(\ell_2)} \leq ||f_A(\theta)||_{M(\ell_2)}$.

Assume next that $A \in L^1(\ell_2)$. We then have that

$$||\sigma_N(f_A)(\theta) - f_A(\theta)||_{M(\ell_2)} = ||\sigma_n(A) - A||_{M(\ell_2)} \to 0.$$

Thus, $\sigma_n(f_A) \to f_A$ uniformly and we obtain that $f_A \in C(\mathbb{T}, M(\ell_2))$.

Assume now that $f_A \in C(\mathbb{T}, M(\ell_2))$. We consider then the $M(\ell_2)$-valued integral

$$f_N(\theta) = \frac{1}{2\pi} \int_{\mathbb{T}} f_A(\theta - t) K_N(t) dt, \quad \theta \in \mathbb{T},$$

where K_N is the Nth Fejer kernel. It is straightforward to see that $f_N - f_A$ converges in $C(\mathbb{T}, M(\ell_2))$ as $N \to \infty$. An easy computation yields that

$$f_N(\theta) = \sum_{n=-N}^{N} \left(1 - \frac{|n|}{N+1} \right) \widehat{f_A}(n) e^{in\theta},$$

where $\widehat{f_A}(n)$ is the nth Fourier coefficient of f_A. To compute the Fourier coefficient $\widehat{f_A}(n)$ we need only to observe that the operation $M \to m_{jk}$ of taking the (j, k)th entry is a bounded linear functional on $M(\ell_2)$. By this we clearly have that $\widehat{f_A}(n) = A_n$. Summing up, we have shown that

$$\lim_{N \to \infty} \sum_{n=-N}^{N} \left(1 - \frac{|n|}{N+1} \right) A_n e^{in\theta} = f_A(\theta)$$

in $C(\mathbb{T}, M(\ell_2))$. For $\theta = 0$ this yields $\sigma_N(A) \to A$ in $M(\ell_2)$. The proof is complete. $\qquad\square$

We also remark that f is continuous on $\mathbb{T}$ if and only if it is continuous at one single point. This is clear by the first assertion of the above theorem.

The next remark is an easy consequence of Fejer's theory.

Remark 3.14. *Let A be a Toeplitz matrix. Then $A \in L^1(\ell_2)$ if and only if it represents a function from $L^1(\mathbb{T})$.*

Now we recall the following well-known result (see [41], Chapter 2). A function F on $\mathbb{T}$ belongs to $L^\infty(\mathbb{T})$ if and only if

$$\sup_n \|\sigma_n(f)\|_{L^\infty(\mathbb{T})} < \infty.$$

We have the following matrix-version of the previous result:

Proposition 3.15. *Let A be an infinite matrix. Then A belongs to $B(\ell_2)$ if and only if*

$$\sup_n \|\sigma_n(A)\|_{B(\ell_2)} < \infty.$$

Proof.   Assume that $\sup_n ||\sigma_n(A)||_{B(\ell_2)} < \infty$. Then, by reasoning as in the proof of Theorem 3.12, we get easily that $A \in B(\ell_2)$.

The converse implication can be proved by using the same arguments as in the proof of Proposition 3.3. □

Proposition 3.16. $A \in M(\ell_2)$ *if and only if*

$$\sup_n ||\sigma_n(A)||_{M(\ell_2)} < \infty.$$

Proof.   Assume that

$$\sup_n ||\sigma_n(A)||_{M(\ell_2)} < \infty$$

and fix an arbitrary $B \in B(\ell_2)$. We then have that

$$||\sigma_n(B * A)||_{B(\ell_2)} \le ||B||_{B(\ell_2)} \sup_n ||\sigma_n(A)||_{M(\ell_2)} < \infty.$$

It follows that $\sigma_n(B * A) \to B * A$ in the weak topology of operators in $B(\ell_2)$. (See the proof of Theorem 3.12.) In particular, $B * A \in B(\ell_2)$. Since B is arbitrary this means that A is a Schur multiplier.

Conversely, let $A \in M(\ell_2)$. Then $\sigma_n(A) = A * \sigma_n(M)$, where $M = (m_i)_{i \in \mathbb{Z}}$ with $m_i = 1$. Thus, by using Bennett's theorem, we find that

$$||\sigma_n(A)||_{M(\ell_2)} \le ||A||_{M(\ell_2)} \cdot ||\sigma_n(M)||_{M(\ell_2)} \le ||A||_{M(\ell_2)}.$$

The proof is complete. □

3.4   A matrix version for functions of bounded variation

Following [75] we introduce now a matrix version of functions with bounded variation and prove the analogue of classical Jordan's theorem on trigonometric series.

We denote by A' the matrix $\sum_{k \in \mathbb{Z}} kA_k$.

Definition 3.17.   *As in the Fourier series framework we say that a matrix A is a matrix of bounded variation if $A' \in M(\ell_2)$.*

The space of all matrices A of bounded variation is denoted by $BV(\ell_2)$ and it is a Banach space endowed with the norm

$$||A||_{BV(\ell_2)} = ||A_0||_{B(\ell_2)} + ||A'||_{M(\ell_2)}.$$

Finally, we say that a matrix A is *absolutely continuous* if $A' \in L^1(\ell_2)$ and $A_0 \in B(\ell_2)$.

Let us remark that by Lemma 3.9 it follows that $||A_k||_{B(\ell_2)} = o\left(\frac{1}{k}\right)$ for an absolutely continuous matrix A.

Moreover, it is easy to see that for a matrix $M \in M(\ell_2)$ we have that $||M_k||_{M(\ell_2)} \le ||M||_{M(\ell_2)}$, for all $k \in \mathbb{Z}$. On the other hand, it is possible that $||M_k||_{M(\ell_2)} \not\to 0$, as $|k| \to \infty$. For instance if M coincides with E, the matrix having only 1 as entries, then obviously

$$\lim_{|k|\to\infty} ||M_k||_{M(\ell_2)} = 1.$$

Next we introduce another interesting subspace of $B(\ell_2)$ by

$$U(\ell_2) = \{A \in C(\ell_2); \text{ such that } ||S_n(A) - A||_{B(\ell_2)} \to 0 \text{ as } n \to \infty\}$$

endowed with the norm

$$||A||_{U(\ell_2)} = \sup_n ||S_n(A)||_{B(\ell_2)}.$$

Obviously, $U(\ell_2) \subset C(\ell_2)$.

Using the well-known example of Du Bois Raymond (see [96]) there exists a Toeplitz matrix from $C(\ell_2)$ which does not belong to $U(\ell_2)$, so that $U(\ell_2)$ *is in fact a proper subspace of* $C(\ell_2)$.

But there are more sophisticated such examples of matrices. For instance we can adapt an example found by Fejer to the framework of infinite matrices which are not Toeplitz matrices.

We sketch in what follows this example:

For any integers n, μ and for all $x \in \mathbb{R}$ we put (see [96]-page 168)

$$\overline{Q}(x,\mu,n) = -\cos(\mu+n)x \sum_{k=1}^{n} \frac{\sin kx}{k}.$$

Since the partial sums of the series $\sin x + \frac{1}{2}\sin 2x + \dots$ are less than a constant C in absolute value, we have that $|\overline{Q}| \le C$, for every x, μ, n.

If we denote by $\overline{Q}(\mu, n)$ the infinite Toeplitz matrix associated to the periodical function $\overline{Q}(x,\mu,n)$, then we get that $||\overline{Q}(\mu,n)||_{B(\ell_2)} \le C$.

Now, for every μ, n, we choose the decreasing sequences $\mathbf{a}_{\mu,n} = \{a_{\mu,n}^l\}_{l \ge 1}$, such that there exist $0 < a < b < \infty$, a, b independent of all indices μ, n, with $a \le a_{\mu,n}^l \le b$, for all $l \ge 1$. Then we consider the matrix $[\mathbf{a}_{\mu,n}]$ defined in [18] as

$$[\mathbf{a}_{\mu,n}] = \begin{pmatrix} a_{\mu,n}^1 & a_{\mu,n}^1 & \cdots & \cdots \\ a_{\mu,n}^1 & a_{\mu,n}^2 & a_{\mu,n}^2 & \cdots \\ a_{\mu,n}^1 & a_{\mu,n}^2 & a_{\mu,n}^3 & \cdots \\ \vdots & \ddots & \ddots & \ddots \end{pmatrix}.$$

Since $\sum_{l=1}^{\infty} |a_{\mu,n}^l - a_{\mu,n}^{l+1}| < \infty$, for all μ, n, we get that $[\mathbf{a}_{\mu,\mathbf{n}}] \in M(\ell_2)$, (see [18]), and $||[\mathbf{a}_{\mu,\mathbf{n}}]||_{M(\ell_2)} \leq b - a$, for all μ, n.

Let $\{n_k\}$, $\{\mu_k\}$ be sets of integers which we shall define in a moment, and let $\alpha_k > 0$, $\alpha_1 + \alpha_2 + \cdots < \infty$. We define the matrices $[\mathbf{a}_{\mu_\mathbf{k},\mathbf{n_k}}]$ as described above. Then the series

$$\sum_{k=1}^{\infty} \alpha_k [\mathbf{a}_{\mu_\mathbf{k},\mathbf{n_k}}] * \overline{Q}(x, \mu_k, n_k)$$

converges uniformly to a continuous matrix, which we denote by $G \in C(\ell_2)$. If $\mu_k + 2n_k < \mu_{k+1}$ $(k = 1, 2, \dots)$, then $\overline{Q}(x, \mu_k, n_k)$ and $\overline{Q}(x, \mu_l, n_l)$ do not overlap for $n \neq l$.

If $\alpha_k = k^{-2}$, $\mu_k = n_k = 2^{k^3}$, then the continuous matrix G defined above has an expansion $G = \sum_{k \in \mathbb{Z}} G_k$ which does not converge in the operator norm. Indeed, the sequence $\mathbf{a}_{\mu_\mathbf{k},\mathbf{n_k}} = \{(a_{\mu_k,n_k}^l)^{-1}\}_{l \geq 1}$ is increasing and also bounded, and, therefore, $[\mathbf{a}_{\mu_\mathbf{k},\mathbf{n_k}}] \in M(\ell_2)$, and its norm in $M(\ell_2)$ is less than $a^{-1} - b^{-1}$. Thus,

$$\alpha_k (\log n_k)/\sqrt{2} < [a^{-1} - b^{-1}] \, ||S_{\mu_k+n_k}(G) - S_{\mu_k-1}(G)||_{B(\ell_2)},$$

which obviously implies that the sequence $S_n(G)$ does not converge in the operator norm.

Next we give the matrix analogue of a well-known Jordan's theorem (see [75]).

Theorem 3.18. *Let $A \in C(\ell_2) \cap BV(\ell_2)$. Then $A \in U(\ell_2)$.*

Proof. Let $A \in C(\ell_2)$. Then it follows that $\sigma_n(A) \to A$. Now let

$$\sigma_{n,k}(A) = \frac{S_n(A) + S_{n+1}(A) + \cdots + S_{n+k-1}(A)}{k}$$

$$= \frac{(n+k)\sigma_{n+k-1}(A) - n\sigma_{n-1}(A)}{k}$$

$$= \left(1 + \frac{n}{k}\right)\sigma_{n+k-1}(A) - \frac{n}{k}\sigma_{n-1}(A).$$

Note that

$$\sigma_{n,k}(A) = S_n(A) + \sum_{\nu+1}^{n+k-1} \left(1 - \frac{\nu - n}{k}\right)(A_\nu + A_{-\nu}).$$

It follows by the first relation that

$$||\sigma_{n,k}(A) - A||_{B(\ell_2)}$$

$$= \left\| \left(1 + \frac{n}{k}\right) \sigma_{n+k-1}(A) - \left(1 + \frac{n}{k}\right) A - \frac{n}{k}\sigma_{n-1}(A) + \frac{n}{k}A \right\|_{B(\ell_2)}$$

$$\leq \left(1 + \frac{n}{k}\right) \|\sigma_{n+k-1}(A) - A\|_{B(\ell_2)} + \frac{n}{k}\|\sigma_{n-1}(A) - A\|_{B(\ell_2)} \to 0.$$

Thus, if $A \in C(\ell_2)$, that is $\sigma_\nu(A) \to A$, as $\nu \to \infty$, and $A \in BV(\ell_2)$, then it follows that $|k|\,\|A_k\|_{B(\ell_2)} \leq C$ for all $k \in \mathbb{Z}$ and, by the second relation above, we get that

$$\|\sigma_{n,k}(A) - S_n(A)\|_{B(\ell_2)} \leq \sum_{|\nu|+1}^{n+k-1} \|A_\nu\|_{B(\ell_2)} \leq C \sum_{n+1}^{n+k-1} \frac{1}{\nu} \leq C\frac{k-1}{n}.$$

Now, if $\epsilon > 0$ and $k = [n\epsilon] + 1$, then $C\frac{k-1}{n} \leq C\epsilon$. Since $\frac{n}{k} < \frac{n}{n\epsilon} < \frac{1}{\epsilon}$ is bounded, it follows that $\|\sigma_n(A) - S_n(A)\|_{B(\ell_2)} \to 0$ and, consequently,

$$\limsup_n \|S_n(A) - A\|_{B(\ell_2)} \leq C\epsilon.$$

Since ϵ can be arbitrarily small it follows that

$$\|S_n(A) - A\|_{B(\ell_2)} \to 0,$$

as $n \to \infty$. The proof is complete. $\square$

3.5   Approximation of infinite matrices by matriceal Haar polynomials

The main goal of this section is to formulate and prove an extension of the approximation theorem of continuous functions by Haar functions (see Theorem **A** in the subsection 3.5.1) to the case with infinite matrices (see Theorem **C** in the subsection 3.5.1). The extension to the matriceal framework is based on one side on the fact that periodical functions belonging to $L^\infty(\mathbb{T})$ may be identified one-to-one with Toeplitz matrices from $B(\ell_2)$ (see Theorem **0** in the subsection 3.5.1), and on the other side on some notions given below; for instance we mention: ms - an unital commutative subalgebra of ℓ^∞, $C(\ell_2)$ the matriceal analogue of the space of all continuous periodical functions $C(\mathbb{T})$, the matriceal Haar polynomials, etc.

After some introductories considerations in subsection 3.5.1, we present some results concerning the ms space, which is important for this generalization, in subsection 3.5.2, and the proof of the main theorem, denoted as Theorem C, is considered in the third subsection.

3.5.1 Introduction

The classical form of Haar's theorem.

Let $\mathbb{T}$ be the one-dimensional torus identified with the interval $[0, 2\pi)$. Now we consider the Haar $L^2(\mathbb{T})$-normalized functions h_k given by $h_0(t) = 1$ for $t \in \mathbb{T}$ and, for $n = 2^k + m$, $k \geq 0$ and $m \in \{0, \ldots, 2^k - 1\}$, by

$$
h_n(t) = \begin{cases}
2^{k/2} & , t \in \Delta_{2m}^{(k+1)}, \\
-2^{k/2} & , t \in \Delta_{2m+1}^{(k+1)}, \\
0 & , t \in \mathbb{T} \setminus \Delta_m^{(k)},
\end{cases}
$$

where $\Delta_m^{(k)} = [\frac{m}{2^k} \cdot 2\pi, \frac{m+1}{2^k} \cdot 2\pi)$.

We can now state the following well-known theorem of approximation of continuous functions on $\mathbb{T}$ (i.e. periodical continuous functions on $[0, 2\pi]$) by means of Haar functions (periodically extended on $\mathbb{R}$) due to Haar (see [37]).

Theorem A. *If f is a continuous function on $\mathbb{T}$ (i.e. if $f \in C(\mathbb{T})$) and if $\epsilon > 0$, then there exists a Haar polynomial of degree $n(\epsilon) \in \mathbb{N}$*

$$
S_n(f) = \sum_{k=0}^{n-1} \alpha_k h_k, \quad \alpha_k \in \mathbb{C},
$$

such that

$$
\|f - S_n(f)\|_{L^\infty(\mathbb{T})} < \epsilon.
$$

Translation of Theorem A to a matriceal framework.

The following result as well as thereafter remark constitute the starting point of the whole theory presented here (see [18]).

Theorem 0. *A Toeplitz matrix $A = (a_k)_{-\infty}^{+\infty}$ belongs to $B(\ell_2)$ if and only if there exists a unique function $f_A \in L^\infty(\mathbb{T})$ whose Fourier coefficients $\widehat{f_A}(n) = \frac{1}{2\pi} \int_0^{2\pi} f(t) e^{-int} dt$ are equal to a_n, for $n \in \mathbb{Z}$ and, moreover,*

$$
\|A\|_{B(\ell_2)} = \|f_A\|_{L^\infty(\mathbb{T})}.
$$

In order to develop the theory we consider in the previous result two different "geometric" directions to be followed.

Model 1: Diagonal matrix.

As we noted already in the introductory chapter, for an infinite matrix $A = (a_{ij})$, and an integer k, possibly negative, we denote by A_k the matrix whose entries $a'_{i,j}$ are given by

$$a'_{i,j} = \begin{cases} a_{i,j} & \text{if } j - i = k, \\ 0 & \text{otherwise.} \end{cases}$$

Then A_k will be called *the kth-diagonal matrix associated to A.*

In the preceding theorem we remark that there is a one-to-one correspondence between A_k and $\widehat{f}_A(k)$ for $A \in B(\ell_2)$ and $f \in L^\infty(\mathbb{T})$.

Consequently, we may imagine $(A_k)_{k \in \mathbb{Z}}$, as the "matriceal Fourier coefficients" associated to the matrix A.

Model 2: Corner matrix.

In the sequel we use another notation, which is more appropriate for our aims. For the entries of the matrix A, we put

$$a_k^l = \begin{cases} a_{l,l+k} \, , & k \geq 0, \ l = 1, 2, 3 \ldots, \\ a_{l-k,l} \, , & k < 0, \ l = 1, 2, 3, \ldots, \end{cases}$$

and denote A sometimes as $A = (a_k^l)_{l \geq 1, \, k \in \mathbb{Z}}$.

Let $A^{(l)} = (b_k^m)_{k \in \mathbb{Z}, \, m \geq 1}$, where $l \in \mathbb{N}^*$, be the matrix given by

$$b_k^m = \begin{cases} a_k^l & \text{if } m = l, \\ 0 & \text{if } m \neq l. \end{cases}$$

We call the matrix $A^{(l)}$, *the lth-corner matrix associated to A.*

Now, if for any corner-matrix $A^{(l)} = (b_k^m)_{k \in \mathbb{Z}, \, m \geq 1}$ we associate a distribution on $\mathbb{T}$, denoted by f_l such that $b_k^l = \widehat{f}_l(k)$, we get, in case $A \in \mathcal{T} \cap B(\ell_2)$, that $f_l = f \in L^\infty(\mathbb{T})$, for all $l \in \mathbb{N}^*$.

Using the models.

a) Model 1.

In this case we recall that A_k plays the role of the "kth Fourier coefficient of the matrix A."

Theorem 0 allows us to write the formula

$$[\mathcal{T} \cap B(\ell_2)]^* = L^\infty(\mathbb{T}),$$

where by $[H]^*$ we denote the image of the space H of matrices by the correspondence $A \to f_A$.

Remark 3.19. *For brevity we write in what follows equations like the previous one in the following manner:*

$$\mathcal{T} \cap B(\ell_2) = L^\infty(\mathbb{T}),$$

$$C(\ell_2) \cap \mathcal{T} = C(\mathbb{T}).$$

b) Model 2.

We can identify the matrix $A = (A^{(l)})_{l \in \mathbb{N}^*}$ with its sequence of associated distributions $\mathbf{f} = (f_l)_{l \in \mathbb{N}^*}$, writing this fact as

$$A = A_{\mathbf{f}}.$$

By Theorem 0 we have the following correspondences:

$f \in L^\infty(\mathbb{T})$ if and only if $A_{\mathbf{f}} \in \mathcal{T} \cap B(\ell_2)$ $\quad \mathbf{f} = (f, f, f, \dots)$

$g \in L^\infty(\mathbb{T})$ if and only if $A_{\mathbf{g}} \in \mathcal{T} \cap B(\ell_2)$ $\quad \mathbf{g} = (g, g, g, \dots).$

Then, of course, it follows that

$fg \in L^\infty(\mathbb{T})$ if and only if $A_{\mathbf{fg}} \in \mathcal{T} \cap B(\ell_2)$ where $\mathbf{fg} = (fg, fg, fg, \dots).$

We recall that the matrix $A = (a_{ij})$ is said to be of n-band type if $a_{ij} = 0$ for $|i - j| > n$.

Having these notions in mind we introduce a commutative product of infinite matrices:

Definition 3.20. *Let $A = A_{\mathbf{f}}$ and $B = A_{\mathbf{g}}$ be two infinite matrices of finite band type. We introduce now the commutative product $\square$ given by*

$$A \square B := A_{\mathbf{fg}}.$$

Remark 3.21. *(1) We mention that in the previous definition since $A = A_{\mathbf{f}}$, and $B = A_{\mathbf{g}}$ are infinite matrices of finite band type it yields that f and g are trigonometric polynomials, and we may consider the product fg.*

2) This product can be defined also for all matrices $A, B \in B(\ell_2)$, but $A \square B$ does then not in general belong to $B(\ell_2)$ as can be easily seen.

3) Of course, if $A_{\mathbf{f}}, A_{\mathbf{g}} \in \mathcal{T} \cap B(\ell_2)$, then it follows that $A_{\mathbf{f}} \square A_{\mathbf{g}} = A_{\mathbf{fg}} \in \mathcal{T} \cap B(\ell_2)$.

We end the presentation of this model by considering an important particular case:

Let $\alpha = (\alpha^1, \alpha^2, \alpha^3, \dots)$ be a sequence of complex numbers and $B = A_{\mathbf{f}} \in B(\ell_2)$, where $\mathbf{f} = (f_1, f_2, \dots)$. Considering α as a sequence of constant functions on $\mathbb{T}$, we get, by Definition 3.20, that

$$A_\alpha \square B = A_{\alpha \mathbf{f}},$$

where $\alpha \mathbf{f} = (\alpha^1 f_1, \alpha^2 f_2, \dots)$.

For brevity we denote $A_\alpha \square B$ by $\alpha \odot B$.

In what follows it will be important to know more about the sequences α satisfying the condition $B \in B(\ell_2) \Rightarrow \alpha \odot B \in B(\ell_2)$.

Actually, the entire next subsection will be devoted to this question, but for the moment, for understanding its implications we will rewrite the operation $\odot$ under a different form.

We associate to any sequence $\alpha = (\alpha^1, \alpha^2, \dots)$, the matrix $[\alpha]$ whose entries $[\alpha]_k^l$ are equal to α^l, for $l \geq 1$ and $k \in \mathbb{Z}$.

Then it is clear that

$$\alpha \odot B = [\alpha] * B.$$

Definition 3.22. *We say that the sequence $\alpha \in ms$ if it has the property that*

$$\alpha \odot B \in B(\ell_2) \quad \forall \, B \in B(\ell_2),$$

or, equivalently,

$$[\alpha] \in M(\ell_2).$$

On ms we consider the norm $\|\alpha\|_{ms} := \|[\alpha]\|_{M(\ell_2)}$. Then ms is a unital commutative Banach algebra with respect to usual multiplication of sequences.

Remark 3.23. *Any constant complex sequence $\alpha = (\alpha, \alpha, \dots)$ belongs to ms.*

In order to get an extension of Haar's theorem we have to find the appropriate analogues in the matrix context. They are summarized below:

The function case	The matrix case
1.　norm $\|\cdot\|_{L^\infty(\mathbb{T})}$	norm $\|\cdot\|_{B(\ell_2)}$
2.　space $C(\mathbb{T})$	space $C(\ell_2)$
3.　multiplication of a function by a scalar	multiplication $\odot$.

The correspondence given by 3. becomes more transparent if we remark that, for $\alpha \in \mathbb{C}$ and for $f \in L^\infty(\mathbb{T})$, denoting by $\widetilde{\alpha}$ the sequence $(\alpha, \alpha, \dots)$, and by $\mathbf{f}$ the constant sequence $(f, f, \dots)$, we get that $\widetilde{\alpha} \odot A_{\mathbf{f}} = [\alpha] * A_{\mathbf{f}} = A_{\alpha f}$.

Denoting by H_k the Toeplitz matrix associated like in Theorem 0 to the Haar function h_k, for $k = 0, 1, 2, \dots$ and by $S_n(\mathbf{f})$ the constant sequence $(S_n(f), S_n(f), \dots)$, where $S_n(f) = \sum_{k=0}^{n-1} \alpha_k h_k$, for $f \in C(\mathbb{T})$, $\alpha_k \in \mathbb{C}$, and $k \in \{0, n-1\}$, we get the following translation of Theorem A in the Toeplitz matrices setting:

Theorem B. *Let $A = A_{\mathbf{f}} \in C(\ell_2)$ be a Toeplitz matrix and let $\epsilon > 0$. Then there is a matriceal polynomial given by*

$$A_{S_n(\mathbf{f})} = \sum_{k=0}^{n-1} \alpha_k H_k = \sum_{k=0}^{n-1} \widetilde{\alpha}_k \odot H_k$$

such that

$$\|A - A_{S_n(\mathbf{f})}\|_{B(\ell_2)} < \epsilon,$$

where $\widetilde{\alpha}_k = (\alpha_k, \alpha_k, \dots)$.

Now it is natural to ask ourselves about the existence of a class of matrices larger than $C(\ell_2) \cap \mathcal{T}$ such that Theorem B still holds.

The aim of our next Theorem is to give an answer to this question. More precisely, we prove the following theorem also formulated in our preface:

Theorem C. *Let $A = (a_k^l)_{l \geq 1,\, k \in \mathbb{Z}}$ be a matrix belonging to $C(\ell_2)$ such that all sequences $\mathbf{a}_k = (a_k^l)_{l \geq 1}$, $k \in \mathbb{Z}$ belong to the class ms.*
Then, for any $\epsilon > 0$ there is an $n_\epsilon \in \mathbb{N}^*$ and sequences $\alpha_k \in ms$, $k \in \{0, \dots, n-1\}$ such that

$$\|A - \sum_{k=0}^{n-1} \alpha_k \odot H_k\|_{B(\ell_2)} < \epsilon.$$

It is also worthwhile to mention the following open problem:

Open problem. *Does Theorem C still hold if the matrix A satisfies only the condition $A \in C(\ell_2)$? If not, what is the best version of Theorem C in this case?*

3.5.2   *About the space ms*

As we remarked in the previous subsection (see also the statement of Theorem C) the space ms plays an important role for our theory and, consequently, it is desirable to know more facts about it.

In this context, we again note that any constant sequence belongs to ms (see Remark 3.23). Our primary goal here is to prove that this algebra is far more rich than that; this richness will quantify the level of extension of the theorem of Haar in the matrix case, since in the functions case, corresponding to Toeplitz matrices, (see Theorem B) the algebra ms is reduced to exactly the constant sequences.

Here is an outlook for this subsection:

We give some sufficient conditions for a sequence to belong to ms, following two complementary ways:

The first one is based on defining a particular algebra pms and showing that pms is intimately connected with ms. (See Proposition 3.25.)

As a consequence we derive properties for ms displaying some necessary and some sufficient conditions for a sequence in order to belong to pms; (see Theorem 3.26) the second approach (Theorem 3.28) is involved with the structure of ms rather than of pms.

For an infinite matrix $A = (a_{ij})_{i\geq 1,\, j\geq 1}$, we define its upper triangular projection $P_T(A)$ as follows:

$$P_T(A) := \begin{cases} a_{i,j} & \text{if } i \leq j, \\ 0 & \text{otherwise.} \end{cases}$$

Definition 3.24. *A sequence* $b = (b_n)_{n\geq 1}$ *belongs to pms if and only if*

$$B := \{b\} = P_T([b]) \in M(\ell_2).$$

Then pms endowed with the norm $||b|| = ||\{b\}||_{M(\ell_2)}$ *becomes a Banach algebra with respect to usual product of sequences.*

Proposition 3.25. *Let* $b = (b_n)_{n\geq 1}$ *be a sequence of complex numbers. Then*

    1. $b \in pms \Rightarrow b \in ms$ *(so $pms \subset ms$.)*

    2. if we write $(b_1, b_2, \ldots, b_n, \ldots) = (b_1, 0, b_3, 0, \ldots) + (0, b_2, 0, b_4, \ldots)$, *or, equivalently,* $b = b^{10} + b^{20}$, *denoting by* $b^1 = (b_1, b_3, \ldots, b_{2n-1}, \ldots)$ *and by* $b^2 = (b_2, b_4, \ldots)$, *we have that* $b^i \in pms \Leftrightarrow b^{i0} \in ms$ *for* $i \in \{1, 2\}$ *and so* $b^i \in pms$, $i \in \{1, 2\}$ $\Rightarrow b \in ms$.

The proof is obvious, so we leave out the details.

We now pass to a study of the algebra *pms*.

We introduce a new method for estimating the norm in the space $B(\ell_2)$.

We associate to every sequence $x = (x_j)_{j \geq 1}$ from $\ell_2(\mathbb{N})$ the function $h(t) = \sum_{j=1}^{\infty} x_j e^{2\pi i j t} \in H_0^2([0,1])$, where $H_0^2([0,1])$ consists of all functions $h : [0,1] \to \mathbb{C}$ from the Hardy space H^2 such that $\int_0^1 h(t)dt = 0$.

If $A = (a_{kj}) \in B(\ell_2)$, we define $\mathcal{L}_k(t) := \sum_{j=1}^{\infty} a_{kj} e^{2\pi i j t} \in H^2([0,1])$.

It follows that

$$\|A\|_{B(\ell_2)} = \sup_{\|h\|_2 \leq 1} \left(\sum_{k=1}^{\infty} \left| \int_0^1 \mathcal{L}_k(t) h(s-t) \, dt \right|^2 \right)^{\frac{1}{2}} < \infty \qquad \text{for any } s. \quad (3.1)$$

See Chapter 2 for a proof of this formula.

Theorem 3.26. *Let* $b = (b_n)_{n \geq 1}$ *be a sequence of complex numbers.*

1) If $(i_n)_{n \geq 1}$ *is a strictly increasing sequence of natural numbers with* $i_1 = 0$, *and* $z_{i_n} := \max_{i_n < k \leq i_{n+1}} |b_k|$, *then there exists a constant* $R > 0$ *such that*

$$\|b\|_{ms} = \|B\|_{M(\ell_2)} \leq R \inf_{(i_n)} \left(\|(z_{i_n})_{n \geq 1}\|_2 + \|(z_{i_n} \ln(i_{n+1} - i_n))_n\|_\infty \right).$$

2) If $b \in pms$, *then* $\sup_{n \geq 1; p \geq 1} \frac{\ln^2 n}{n} \sum_{k=p}^{n+p} |b_k|^2 < \infty$.

3) If $(|b_k|)_{k \geq 1}$ *is a decreasing sequence, then* $b \in pms$ *if and only if* $|b_k| = \mathcal{O}\left(\frac{1}{\ln k}\right)$.

Proof. 1) Let $A \in B(\ell_2)$ and $x \in \ell_2(\mathbb{N})$. By using the relation (3.1) for $B \star A$ it follows that there exists $R_1 > 0$ such that

$$\|(B \star A)x\|_2^2 \leq R_1 \sum_{k=1}^{\infty} |b_k|^2 \left| \int_0^1 \mathcal{L}_k(t) \left(h - S_{k-1}(h)(-t) \right) dt \right|^2,$$

where $S_k(h)$ is the Fourier partial sum of order k (i.e. if D_k is the Dirichlet kernel, then $S_k(h)(t) = (h \star D_k)(t)$ is the convolution of h and D_k).

Therefore, we have that

$$\|(B \star A)x\|_2^2 \leq 2 \sum_{k=1}^{\infty} |b_k|^2 \left[\left| \int_0^1 \mathcal{L}_k(t) S_{k-1}(h)(-t) dt \right|^2 + \left| \int_0^1 \mathcal{L}_k(t) h(-t) dt \right|^2 \right]$$

$$\leq 2 \sum_{k=1}^{\infty} |b_k|^2 \left| \int_0^1 \mathcal{L}_k(t) S_{k-1}(h)(-t) dt \right|^2 + 2\|b\|_\infty^2 \sum_{k=1}^{\infty} \left| \int_0^1 \mathcal{L}_k(t) h(-t) dt \right|^2$$

$$\leq 2 \sum_{k=1}^{\infty} |b_k|^2 \left| \int_0^1 \mathcal{L}_k(t) S_{k-1}(h)(-t) dt \right|^2 + 2\|b\|_\infty^2 \|A\|_{B(\ell_2)}^2 \|h\|_2^2.$$

Let $(i_n)_{n\geq 1}$ be a strictly increasing sequence of natural numbers such that $i_1 = 0$.

Then we have that

$$||b||_\infty^2 \leq ||\,(z_{i_n})_{n\geq 1}\,||_2^2 \qquad (3.2)$$

and

$$\sum_{k=1}^{\infty} |b_k|^2 \left| \int_0^1 \mathcal{L}_k(t) S_{k-1}(h)(-t)\, dt \right|^2$$

$$= \sum_{n=1}^{\infty} \sum_{k=i_n+1}^{i_{n+1}} |b_k|^2 \left| \int_0^1 \mathcal{L}_k(t) S_{k-1}(h)(-t)\, dt \right|^2$$

$$\leq 2 \left(\sum_{n=1}^{\infty} \sum_{k=i_n+1}^{i_{n+1}} |b_k|^2 \left| \int_0^1 \mathcal{L}_k(t) S_{i_n}(h)(-t)\, dt \right|^2 \right)$$

$$+ 2 \left(\sum_{n=1}^{\infty} \sum_{k=i_n+1}^{i_{n+1}} |b_k|^2 \left| \int_0^1 \mathcal{L}_k(t) (S_{k-1} - S_{i_n})(h)(-t)\, dt \right|^2 \right)$$

$$\leq 2 \sum_{n=1}^{\infty} z_{i_n}^2 \left(\sum_{k=i_n+1}^{i_{n+1}} \left| \int_0^1 \mathcal{L}_k(t) S_{i_n}(h)(-t)\, dt \right|^2 \right) +$$

$$2 \sum_{n=1}^{\infty} z_{i_n}^2 \left(\sum_{k=i_n+1}^{i_{n+1}} \left| \int_0^1 \mathcal{L}_k(t)(S_k - S_{i_n})(h)(-t)\, dt \right|^2 \right).$$

Moreover, by using the formula (3.1), we get that

$$\sum_{k=i_n+1}^{i_{n+1}} \left| \int_0^1 \mathcal{L}_k(t) S_{i_n}(h)(-t)\, dt \right|^2 \leq ||A||_{B(\ell_2)}^2 \, ||h||_2^2$$

and

$$\sum_{k=i_n+1}^{i_{n+1}} \left| \int_0^1 \mathcal{L}_k(t)(S_{k-1} - S_{i_n})(h)(-t)\, dt \right|^2$$

$$\sim \sum_{k=1}^{i_{n+1}-i_n} \left| \int_0^1 D_{k-1}(t)\{[\mathcal{L}_{k+i_n} \star (S_{i_{n+1}} - S_{i_n})(h)](-t) e^{2\pi i_n it}\}\, dt \right|^2$$

$$\leq \sup_{1 \leq k \leq i_{n+1}-i_n} \|D_{k-1}\|_{L^1(0,1)}$$

$$\times \sum_{k=1}^{i_{n+1}-i_n} \int_0^1 |D_{k-1}(t)| \left|\left(\mathcal{L}_{k+i_n} \star \left(S_{i_{n+1}} - S_{i_n}\right)(h)\right)(-s)\right|^2 ds$$

$$\leq C \log\left(i_{n+1}-i_n\right) \left(\int_0^1 \sup_{1 \leq k \leq i_{n+1}-i_n} |D_{k-1}(s)| ds\right)$$

$$\times \left\|\sum_{k=1}^{i_{n+1}-i_n} \left|\left(\mathcal{L}_{k+i_n} \star \left(S_{i_{n+1}} - S_{i_n}\right)(h)\right)(\cdot)\right|^2\right\|_{L^\infty}$$

$$\leq C' \|A\|_{B(\ell_2)}^2 \|(S_{i_{n+1}} - S_{i_n})(h)\|_2^2 |\ln(i_{n+1}-i_n)|^2.$$

Thus, using (3.2), we get that

$$\|(A \star B)x\|_{B(\ell_2)} \leq R \|A\|_{B(\ell_2)} \|h\|_2 (\|(z_{i_n})_{n \geq 1}\|_2 + \|z_{i_n} \ln(i_{n+1}-i_n)\|_\infty).$$

2) Let $B \in M(\ell_2)$.

Taking $A \in \mathcal{T} \cap B(\ell_2)$ such that $a_j^l = \frac{1}{j}$ for all $j \in \mathbb{Z} \setminus \{0\}$ and for all $l \in \mathbb{N}^*$ and $a_0^l = 0$ for all $l \in \mathbb{N}$, we obtain that $\tilde{B} \star \tilde{A} \in B(\ell_2)$, where

$$\tilde{B} := \begin{pmatrix} b_1 & b_1 & b_1 & b_1 & \cdots \\ b_2 & b_2 & b_2 & b_2 & \cdots \\ \cdots & \cdots & \cdots & \cdots & \cdots \\ b_n & b_n & b_n & b_n & \cdots \\ \cdots & \cdots & \cdots & \cdots & \cdots \end{pmatrix} \text{ and } \tilde{A} := \begin{pmatrix} 1 & \frac{1}{2} & \frac{1}{3} & \cdots \\ \frac{1}{2} & 1 & \frac{1}{2} & \cdots \\ \frac{1}{3} & \frac{1}{2} & 1 & \cdots \\ \cdots & \cdots & \cdots & \cdots \end{pmatrix}.$$

Letting $x_p^n = (x_k)_{k \geq 1}$ with $x_k = \begin{cases} 1 \text{ if } k \in \{p, \ldots, n+p\} \\ 0 \text{ otherwise} \end{cases}$, where $p, n \in \mathbb{N}^*$ are fixed, we get that

$$\ln^2(n+1) \sum_{k=p}^{n+p} |b_k|^2 \leq C \left\|\left(\tilde{B} \star \tilde{A}\right) x_p^n\right\|_2^2 \leq C(n+1).$$

Hence,

$$\sup_{n \geq 1; p \geq 1} \frac{\ln^2(n+1)}{n+1} \sum_{k=p}^{n+p} |b_k|^2 < \infty.$$

3) Let $(|b_k|)_{k \geq 1}$ be a decreasing sequence. Then, according to 2), we get that $|b_n| = \mathcal{O}\left(\frac{1}{\ln n}\right)$.

Conversely, defining $r = (r_n)_{n \geq 1}$ with $r_n = \frac{1}{\ln(n+1)}$ for all $n \geq 1$, we have that

$$||B||_{M(\ell_2)} = ||\{b\}||_{M(\ell_2)} \leq C||\{r\}||_{M(\ell_2)}.$$

By choosing $i_n = 2^n$ for all $n \geq 2$ and $i_1 = 0$, it follows, by 1), that $z_{i_n} = r_{2^n+1} \sim \frac{1}{n}$.

Consequently,

$$||\{r\}||_{M(\ell_2)} \leq R \left\{ \left\|\left(\frac{1}{n}\right)_{n \geq 1}\right\|_2 + \left\|\left(\frac{1}{n}\ln 2^n\right)_{n \geq 1}\right\|_\infty \right\} < \infty$$

that is $B \in M(\ell_2)$. The proof is complete.   □

Observe that results like Theorem 7.1 or Theorem 8.6 [11] cannot be applied in our situation.

Remark 3.27. *From the previous results we deduce that*

$$\ell_2(\mathbb{N}) \subset ms \subset \ell_\infty(\mathbb{N})$$

and that $\{(b_n)_{n \geq 1} \mid |b_n| = \mathcal{O}\left(\frac{1}{\ln n}\right)\} \subset ms$, with proper inclusions.

Next we change the view and derive another set of sufficients conditions in order that $b \in ms$. These results use the estimate of the absolute value of differences of sequence's terms rather than the absolute value of the terms themselves.

Theorem 3.28. *Let $b = (b_n)_{n \geq 1}$ be a sequence of complex numbers.*

(1) If $\sup_{n \geq 1} \left(\sum_{j=1}^n |b_j - b_n|^2\right) < \infty$, then $b \in ms$,

(2) If $||b||_{BV(\mathbb{N})} = |b_1| + \sum_{n=1}^\infty |b_{n+1} - b_n| < \infty$, then $b \in ms$.

Proof. 1. We use the following result from [11]

A matrix $M \in M(\ell_2)$ iff there exists a $P \in B(\ell_2, \ell_\infty)$ and $Q \in B(l_1, \ell_2)$ such that

$$M = PQ \quad and \quad ||M||_{M(\ell_2)} \leq ||P||_{2,\infty}||Q|_{1,2}.$$

We recall that if $Q = (q_{jk})$, $j \geq 1$, $k \geq 1$ and $P = (p_{jk})$, $j \geq 1$, $k \geq 1$, then

$$||Q||_{1,2} = \sup_{k \geq 1}\left(\sum_{j \geq 1}|q_{jk}|^2\right)^{\frac{1}{2}} \quad \text{and} \quad ||P||_{2,\infty} = \sup_{j \geq 1}\left(\sum_{k \geq 1}|p_{jk}|^2\right)^{\frac{1}{2}}.$$

Let $[b] = B_b + C_b$, where

$$B_b = \begin{pmatrix} b_1 & b_2 & b_3 & \cdots \\ b_1 & b_2 & b_3 & \cdots \\ b_1 & b_2 & b_3 & \cdots \\ \vdots & \vdots & \vdots & \ddots \end{pmatrix},$$

and

$$C_b = \begin{pmatrix} 0 & b_1 - b_2 & b_1 - b_3 & \cdots \\ 0 & 0 & b_2 - b_3 & \cdots \\ 0 & 0 & 0 & \cdots \\ \vdots & \vdots & \vdots & \ddots \end{pmatrix}.$$

$B_b \in M(\ell_2) \, \forall b \in \ell_\infty$, and, since $\|C_b\|_{1,2} = \sup_n \left(\sum_{j=1}^n |b_j - b_n|^2 \right)^{\frac{1}{2}} < \infty$, by Bennett's theorem it follows that $C_b \in M(\ell_2)$.

2. If $A \in B(\ell_2)$, $(\mathcal{L}_k)_{k \geq 1}$, $h \in H^2([0,1])$, $x \in l^2(\mathbb{N})$ and $h_k = k \star D_k$ are as before (3.1) and defining $f(t) = \sum_{j=1}^\infty b_j e^{2\pi i j t}$ (in the sense of distributions) we get that

$$\| ([b] \star A) x \|_2^2 =$$

$$\sum_{k=1}^\infty \left| \int_0^1 f(t) \, (\mathcal{L}_k \star h_k)(-t) dt + \int_0^1 f(t) \, (\mathcal{L}_k \star (h - h_k))(0) e^{-2\pi i k t} dt \right|^2 \leq$$

$$2 \sum_{k=1}^\infty \left| \int_0^1 g_k(s) \, (\mathcal{L}_k \star h)(-s) ds \right|^2 + 2 \|b\|_\infty^2 \|A\|_{B(\ell_2)}^2 \|x\|_2^2,$$

where

$$g_k(s) = \sum_{j=1}^k \left(\hat{f}(k) - \hat{f}(j) \right) e^{2\pi i j s} = \sum_{j=1}^{k-1} \left(\hat{f}(j+1) - \hat{f}(j) \right) D_j(s).$$

But

$$\sum_{k=1}^\infty \left| \int_0^1 g_k(s) \, (\mathcal{L}_k \star h)(-s) ds \right|^2 \leq$$

$$\left(\sum_{j=1}^\infty \left| \hat{f}(j+1) - \hat{f}(j) \right| \right) \sum_{j=1}^\infty \left| \hat{f}(j+1) - \hat{f}(j) \right| \sum_{k=2}^\infty |(h_j \star \mathcal{L}_k)(0)|^2 \leq$$

$$\left(\sum_{j=1}^\infty \left| \hat{f}(j+1) - \hat{f}(j) \right| \right)^2 \|h\|_2^2 \|A\|_{B(\ell_2)}^2,$$

so that $\| ([b] \star A) x \|_2 \leq C \|A\|_{B(\ell_2)} \|x\|_2 \left(\sum_{j=1}^\infty |b_{j+1} - b_j| + \|b\|_\infty \right)$, that is $[b] \in M(\ell_2)$. $\qquad \square$

3.5.3   *Extension of Haar's theorem*

As we announced this section is dedicated to prove the generalized Haar theorem - see Theorem C from subsection 3.5.1. We start our exposition by introducing a useful vector space $E(\ell_2)$. After that, we define the notion of generalized scalar product for matrices which allows us to give a more useful form for $E(\ell_2)$ and also to see some similarities with what happens in the function case.

Proposition 3.29 is useful to identify the constraints in the definition of $E(\ell_2)$ and also to point out some of the difficulties in this theory.

Finally, we define the space $C_r(\ell_2)$ and prove that this space admits a Schauder type decomposition (see Proposition 3.30, Theorem 3.31). In this more general frame Theorem C will follow as a Corollary.

We define the vector space $E(\ell_2)$ as follows:

$E(\ell_2) := \{A \in B(\ell_2); A = \sum_{k=0}^{n} \alpha_k \odot H_k$, such that $\alpha_k \odot H_k \in B(\ell_2)$ for all $0 \le k \le n, \ n \in \mathbb{N}\}$, where $\alpha_k \in \ell_\infty$, and $\alpha_k \odot H_k = [\alpha_k] \star H_k$.

We introduce a *generalized scalar product of matrices* $< A, B >$ for $A = A_{\mathbf{f}}$ and $B = B_{\mathbf{g}}$, where $\mathbf{f} = (f_1, f_2, \dots)$ and $\mathbf{g} = (g_1, g_2, \dots)$, in the following way:

$$< A, B > = (< f_1, g_1 >, < f_2, g_2 >, \dots).$$

We say that a family $(\Phi_k)_{k \in \mathbb{N}}$ is *an orthonormal system* if the following orthogonality relations hold: $\langle \Phi_k, \Phi_l \rangle = \mathbf{0} \in \ell_\infty$ for $k \ne l$ and $\langle \Phi_k, \Phi_k \rangle = 1 \in \ell_\infty$ for all $k \in \mathbb{N}^*$.

By the orthogonality of the system $(H_k)_k$ we deduce that $A \in E(\ell_2)$ implies that $A = \sum_{l=1}^{n} < A, H_k > \odot H_l \in E(\ell_2)$.

Therefore,

$$E(\ell_2) = \{A \in B(\ell_2); A = \sum_{l=1}^{n} \langle A, H_l \rangle \odot H_l, \text{ such that}$$
$$\langle A, H_l \rangle \odot H_l \in B(\ell_2) \text{ for all } l \le n, \ n \in \mathbb{N}^*\}.$$

Proposition 3.29. 1) There is $A \in B(\ell_2)$ such that $\langle A, H_1 \rangle \in \ell_\infty$ and $\langle A, H_1 \rangle \odot H_1 \notin B(\ell_2)$.

2) If $0 < p \le 2$ and $A \in S_p$, where S_p is the Schatten class of order p, (see, for instance, [94] for the definition of a Schatten class), then $[< A, H_k >] \in M(\ell_2)$, which, in turn, implies that $< A, H_k > \odot H_k \in B(\ell_2)$ for any $k \in \mathbb{N}^*$.

Proof.   1) Let $A = A_1$ with $a_1^{2k-1} = 1$, $a_1^{2k} = 0$ for $k \in \mathbb{N}^*$ and $a_k^l = 0$, if $k \ne 1$ and $k \in \mathbb{N}^*$. Then $\langle A, H_1 \rangle = (x_1, 0, x_1, 0, \dots) \in \ell_\infty$, where x_1 is some constant.

Hence $\langle A, H_1 \rangle \odot H_1 = \dfrac{2i}{\pi} x_1 B$, where

$$B = \begin{pmatrix} 0 & 1 & 0 & \frac{1}{3} & 0 & \frac{1}{5} & \cdots \\ -1 & 0 & 0 & 0 & 0 & 0 & \ddots \\ 0 & 0 & 0 & 1 & 0 & \frac{1}{3} & \ddots \\ -\frac{1}{3} & 0 & -1 & 0 & 0 & 0 & \ddots \\ 0 & 0 & 0 & 0 & 0 & 1 & \ddots \\ -\frac{1}{5} & 0 & -\frac{1}{3} & 0 & -1 & 0 & \ddots \\ \vdots & \ddots & \ddots & \ddots & \ddots & \ddots & \ddots \end{pmatrix}.$$

But, $\|I - P_T(B)\|_{B(\ell_2)}$ is infinite and $\|I - P_T(B)\|_{B(\ell_2)} \le \|B\|_{B(\ell_2)}$, where I is the unit for the usual non-commutative multiplication of infinite matrices.

This result is not surprising, since, by using Proposition 3.25 and Theorem 3.26, we obtain that

$$(x_1, 0, x_1, 0, \dots) \in ms \iff x_1 = 0.$$

2) Let $p \le 2$. By [94], it yields that any $A \in B(\ell_2)$ belongs to S_p if and only if, for any orthonormal basis (e_k) in ℓ_2, we have that $\sum_k \|Ae_k\|^p < \infty$, hence $\sum_{k=1}^{\infty} \left(\sum_{j=1}^{\infty} |a_{kj}|^2 \right)^{\frac{p}{2}} < \infty$, $\sum_{j=1}^{\infty} \left(\sum_{k=1}^{\infty} |a_{kj}|^2 \right)^{\frac{p}{2}} < \infty$.

Thus, by using Cauchy-Schwarz inequality and the above inequalities, we get that

$$\| < A, H_k > \|_p^p \le C \|H_k\|_{B(\ell_2)}^p \|A\|_{S_p}^p < \infty$$

for some constant $C > 0$.

By Remark 3.27 it follows that $[< A, H_k >] \in M(\ell_2)$. The last implication is now obvious. The proof is complete. $\qquad\square$

Observe also that there exists an $A \in B(\ell_2)$ such that $\langle A, H_k \rangle \odot H_k \in B(\ell_2)$, for all $k \in \mathbb{N}$, but for a $k_0 \in \mathbb{N}$, we find that $\langle A, H_{k_0} \rangle \notin ms$. Indeed $A = A_0 = (a_n)_{n \ge 1} \in \ell_\infty \setminus ms$ gives the answer to the above problem for $k_0 = 0$.

Therefore, in the definition of $E(\ell_2)$, we prefer the weaker condition $< A, H_k > \odot H_k \in B(\ell_2)$ for all k rather than $< A, H_k > \in ms$ for all k.

On the ms-module $E(\ell_2)$ we consider the norm

$$\|\|A\|\| := \sup_{m \le n} \left\| \sum_{k=0}^{m} < A, H_k > \odot H_k \right\|_{B(\ell_2)}.$$

Since $E(\ell_2) \cap \mathcal{T}$ can be identified with $E_d([0,1])$, the space of all dyadic step functions, whose completion with respect to the supremum norm is equal to the space of all countable piecewise continuous functions with discontinuities at dyadic points of $[0,1]$. This space is denoted by $C_r([0,1])$, and we call $C_r(\ell_2)$ the completion of $(E(\ell_2), ||| \cdot |||)$.

In what follows we give some known classes of matrices which are embedded in $C_r(\ell_2)$.

Examples. 1) Obviously all Toeplitz matrices, associated to functions from $C_r([0,1])$ belong to $C_r(\ell_2)$.

2) The Hilbert-Schmidt matrices $A = \left(a_j^l \right)$, $j \in \mathbb{Z}$, $l \geq 1$, with $\|A\|_{HS} = \left(\sum_{j=1}^{\infty} \sum_{l=1}^{\infty} \left| a_j^l \right|^2 \right)^{\frac{1}{2}} < \infty$ belong to $C_r(\ell_2)$ and $\|A\|_{C_r(\ell_2)} \leq \sqrt{2} \|A\|_{HS}$.

We denote by $\overset{\vee}{g}(t) = g(-t)$ and $P_l \left(\sum_{j=-\infty}^{\infty} a_j e^{2\pi ijt} \right) = \sum_{j=l}^{\infty} a_j e^{2\pi ijt}$. Then, by the Fubini theorem and the Cauchy-Schwarz inequality, we get that

$$\|P_{\mathcal{T}}(S_n(A))\|_{B(\ell_2)}^2 = \sup_{\|g\|_{H^2([0,1])} \leq 1} \sum_{l=1}^{\infty} \left| \int_0^1 S_n(f_l)(P_l g)(-t) dt \right|^2$$

$$= \sup_{\|g\|_{H^2([0,1])} \leq 1} \sum_{l=1}^{\infty} \left| \int_0^1 f_l S_n \left(P_l \overset{\vee}{g} \right)(-t) dt \right|^2$$

$$\leq \|A\|_{HS}^2 \cdot \sup_{\|g\|_{H^2([0,1])} \leq 1} \left\| P_l \overset{\vee}{g} \right\|_{L^2}^2 = \|A\|_{HS}^2.$$

Hence, $\|A\|_{C_r(\ell_2)} \leq \sqrt{2} \|A\|_{HS}$.

3) Let A be a diagonal matrix having as non-zero entries the elements of the sequence $\alpha = (\alpha_i)_{i \geq 1} \in ms$. Then $A \in C_r(\ell_2)$ and $\|A\|_{C_r(\ell_2)} \leq \|\alpha\|_{ms}$.

The proof is straightforward using the trivial observations that ms is an algebra with respect to usual multiplication and $C_r(\ell_2)$ is a ms-module with $||| \alpha \odot X ||| \leq \|\alpha\|_{ms} \cdot ||| X |||$.

4) If $A = \left(a_j^l \right)_{j \in \mathbb{Z}, l \geq 1}$ is such that $\sum_{j=-\infty}^{\infty} \|a_j\|_{ms} < \infty$, where $a_j := \left(a_j^l \right)_{l \geq 1}$, then $\|A\|_{C_r(\ell_2)} \leq \sum_{j=-\infty}^{\infty} \|a_j\|_{ms}$ and $A \in C_r(\ell_2)$.

This statement follows easily from 3).

5) If A is the main diagonal matrix having as non-zero entries the elements a_j with $(a_j)_j \in \ell_\infty$, then $A \in C_r(\ell_2)$ and $\|A\|_{B(\ell_2)} = \|A\|_{C_r(\ell_2)}$. (Note that $(a_j)_j$ may not belong to ms.)

Proposition 3.30. *If the sequence of matrices* $(A^n)_{n \geq 1}$ *is a Cauchy sequence of* $E(\ell_2)$ *with respect to norm* $||| \cdot |||$, *then* $\langle A^n, H_k \rangle \odot H_k$ *converges to* some $\alpha_k \odot H_k$ in this norm. Moreover, $\alpha_k \odot H_k \in B(\ell_2)$ and $\langle A^n, H_k \rangle \underset{n}{\to} \alpha_k$ in ℓ_∞.

Proof. *Step I.* We first prove that

$$\|\langle A, H_k \rangle\|_{\ell_\infty} \le 2\,\|A\|_{B(\ell_2)} \qquad \text{for all } k \in \mathbb{N} \text{ and } A \in B(\ell_2). \qquad (3.3)$$

If $A = A_{\mathbf{f}}$, where $\mathbf{f} = (f_1, f_2, \dots)$, and $Q_l A$ is the matrix with entries

$$[Q_l A]_j^k = \begin{cases} a_j^l & k = l, j \in \mathbb{Z} \\ 0 & k \neq l. \end{cases},$$

then, by the Cauchy-Schwarz inequality, it follows that

$$\|\langle A, H_k \rangle\|_{\ell_\infty} = \|(\langle f_l, h_k \rangle)_l\|_\infty \le \sup_{l \in \mathbb{N}^*} \|f_l\|_{L^2} = \sup_{l \in \mathbb{N}^*} \left(\sum_{j=-\infty}^{\infty} |a_j^l|^2 \right)^{\frac{1}{2}}$$

$$\le \sqrt{2} \sup_{l \in \mathbb{N}^*} \|Q_l A\|_{B(\ell_2)} \le 2\,\|A\|_{B(\ell_2)}.$$

Step II. Let now $(A^n)_{n \ge 1}$ be a Cauchy sequence in $E(\ell_2)$. Then, for a fixed $k \in \mathbb{N}$, we have that $\langle A^n, H_k \rangle \underset{n}{\to} \alpha_k$ in ℓ_∞.

Indeed, using (3.3) and the fact that $\|A\|_{B(\ell_2)} \le |||A|||$, the statement follows by Step I.

Step III. $(\langle A^n, H_k \rangle \odot H_k)_{n \ge 1}$ is a Cauchy sequence of $E(\ell_2)$ for all k and for a Cauchy sequence $(A_{n \ge 1}^n)$ in $E(\ell_2)$. Hence $\langle A^n, H_k \rangle \odot H_k \underset{n}{\to} B^k \in C_r(\ell_2)$ in the norm $||| \cdot |||$. Thus, by (3.3), it follows that $\lim_n \|\langle A^n, H_k \rangle - \langle B^k, H_k \rangle\|_{\ell_\infty} = 0$, and by Step II it follows that $\alpha_k = \langle B^k, H_k \rangle$.

Step IV. If we show that $B^k = \langle B^k, H_k \rangle \odot H_k$, then Proposition 3.30 is proved. But by Step III we have that $\langle A^n, H_k \rangle \odot H_k \underset{n}{\to} B^k$ in $B(\ell_2)$. Then the entries of the matrices $\langle A^n, H_k \rangle \odot H_k$ converge with respect to n to the corresponding entry of the matrix B^k. Moreover, according to Step I, $\langle A^n, H_k \rangle \underset{n}{\to} \langle B^k, H_k \rangle$ in ℓ_∞ and, hence, it follows that $\langle B^k, H_k \rangle \odot H_k = B^k$. The proof is complete. $\square$

We use Proposition 3.30 in order to prove the existence of some kind of Schauder basis in $C_r(\ell_2)$ given by the sequence $(H_k)_{k \ge 0}$.

More specifically, we have the following result:

Theorem 3.31. *Let $A \in C_r(\ell_2)$. Then we have the decomposition*

$$A = \sum_{k=0}^{\infty} \langle A, H_k \rangle \odot H_k,$$

in the norm $||| \cdot |||$.

Proof. Let $A \in C_r(\ell_2)$. Then there exists a Cauchy sequence $A^n \in E(\ell_2)$ such that $A = \lim_n A^n$.

By Proposition 3.30 we obtain that $\lim_{n \to \infty} |||\langle A^n, H_k \rangle \odot H_k - \alpha_k \odot H_k||| = 0$ for all $k \geq 0$. Let $\varepsilon > 0$. Therefore, there exists $n_\varepsilon \geq 0$ such that for all $n \geq n_\varepsilon$, and $k > j$, we get that

$$||| \sum_{i=j}^{k} \langle A^n, H_i \rangle \odot H_i - \sum_{i=j}^{k} \alpha_i \odot H_i ||| \leq \limsup_{m \to \infty} ||| \sum_{i=j}^{k} \langle A^n, H_i \rangle \odot H_i - \quad (3.4)$$

$$\sum_{i=j}^{k} \langle A^m, H_i \rangle \odot H_i ||| + \lim_{m \to \infty} \sum_{i=j}^{k} ||| \langle A^m, H_i \rangle \odot H_i - \alpha_i \odot H_i ||| \leq \varepsilon.$$

By the orthogonality relations satisfied by the sequence $(H_k)_k$ and using (3.4), we find that there exists a number l_ε such that $||| \sum_{i=j}^{k} \alpha_i \odot H_i ||| < \varepsilon$ for all $k > j > l_\varepsilon$.

Thus, $\sum_{i=0}^{\infty} \alpha_i \odot H_i = B \in C_r(\ell_2)$. But, by taking $j = 0$ and $k \geq \max(k(n), l_\varepsilon)$, where $\sum_{i=0}^{k(n)} \langle A^n, H_i \rangle \odot H_i = A^n$, in (3.4), we find that, for all $\varepsilon > 0$, and for all $n \geq n_\varepsilon$, $|||A^n - \sum_{i=0}^{k(n)} \alpha_i \odot H_i||| < \varepsilon$.

Thus $A = B = \sum_{i=0}^{\infty} \alpha_i \odot H_i$ and, using the orthogonality relations satisfied by $(H_k)_k$ and the fact that the operator $A \to \langle A, H_i \rangle : C_r(\ell_2) \to \ell_\infty$ is continuous, we conclude that $A = \sum_{i=0}^{\infty} < A, H_i > \odot H_i$ the series being convergent in $C_r(\ell_2)$. The proof is complete. $\square$

In particular, we get the following extension of Haar's theorem for matrices:

Corollary 3.32. *Let $A \in C_r(\ell_2)$. Then $A = \sum_{k=0}^{\infty} \langle A, H_k \rangle \odot H_k$, in the norm of $B(\ell_2)$.*

Of course there exists $A \in C(\ell_2) \backslash C_r(\ell_2)$, for instance A is the diagonal matrix A_1 given by the sequence $(a_n)_{n \geq 1}$, where $a_{2n-1} = 1$ and $a_{2n} = 0$ for all $n = 1, 2, 3, \ldots$

Proof of Theorem C. Let A be an infinite matrix as in Theorem C and let $\epsilon > 0$. Since $A \in C(\ell_2)$ there is $k \in \mathbb{N}$ such that $||\sigma_k(A) - A||_{B(\ell_2)} < \frac{\epsilon}{2}$. Then, by hypothesis and by Example 4, it follows that $\sigma_k(A) \in C_r(\ell_2)$, and consequently, by Theorem 3.31, there is a Haar polynomial $\sum_{i=0}^{n-1} \alpha_i \odot H_i$ such that $||\sigma_k(A) - \sum_{i=0}^{n-1} \alpha_i \odot H_i||_{B(\ell_2)} < \frac{\epsilon}{2}$. The proof is complete. $\square$

3.6 Lipschitz spaces of matrices; a characterization

Let $1 \leq p < \infty$ and let $A \in C_p$. We define

$$\omega_p(\delta) = \omega_p(\delta; A) := \sup_{0 < h \leq \delta} \left\{ \int_0^{2\pi} ||A(x+h) - A(x)||_{C_p}^p \, dx \right\}^{1/p},$$

where $\omega_p(\delta)$ is called *the p-modulus of continuity of A*. If $\omega_p(\delta) \leq M\delta^\alpha$, where $0 < \alpha < 1$, then, we write $A \in Lip(\alpha, p; \ell^2)$ and we say that A *satisfies a Lipschitz condition in the p-metric*.

Now we indicate a relation between the differentiability properties of the matrix A and the behaviour of its "Fourier coefficients" A_k.

Theorem 3.33. *Let* $A \in C_2$.
If

$$A \in Lip(\alpha, 2; \ell^2) \quad (0 < \alpha < 1), \tag{3.5}$$

then

$$\sum_{|k| \geq n} ||A_k||_{C_2}^2 \leq C_1 \frac{1}{n^{2\alpha}} \quad (n \geq 1), \tag{3.6}$$

and conversely.

Proof. Since

$$A\left(x + \frac{\pi h}{4}\right) - A\left(x - \frac{\pi h}{4}\right) \sim 2i \sum_{k=-\infty}^{\infty} e^{ikx} \sin \frac{k\pi h}{4} A_k,$$

then

$$\frac{1}{2\pi} \int_0^{2\pi} ||A\left(x + \frac{\pi h}{4}\right) - A\left(x - \frac{\pi h}{4}\right)||_{C_2}^2 dx = 4 \sum_{k=-\infty}^{\infty} ||A_k||_{C_2}^2 \sin^2 \frac{k\pi h}{4}. \tag{3.7}$$

If $A \in Lip(\alpha, 2; \ell^2)$ $(0 < \alpha < 1)$, then, by (3.7), it follows that

$$4 \sum_{k=-\infty}^{\infty} ||A_k||_{C_2}^2 \sin^2 \frac{k\pi h}{4} \leq 2Ch^{2\alpha},$$

where C does not depend on h. Thus,

$$\sum_{\frac{1}{h} \leq |k| < \frac{2}{h}} ||A_k||_{C_2}^2 \leq Ch^{2\alpha},$$

or, equivalently

$$\sum_{\frac{2^r}{h} \leq |k| < \frac{2^{r+1}}{h}} ||A_k||_{C_2}^2 \leq C \frac{h^{2\alpha}}{2^{2r\alpha}} \quad (r = 0, 1, 2, \ldots).$$

Adding these inequalities we find that

$$\sum_{|k|\geq\frac{1}{h}} ||A_k||^2_{C_2} \leq C\frac{4^\alpha}{4^\alpha - 1}h^{2\alpha},$$

and, consequently, (3.5) implies (3.6).

Assume now (3.6). Put in (3.6) $n = \left[\frac{1}{h}\right] + 1$. Then, in view of (3.7), it follows that

$$\frac{1}{2\pi}\int_0^{2\pi} ||A(x+\frac{\pi h}{4}) - A(x - \frac{\pi h}{4})||^2_{C_2}dx \leq 4C_1 h^{2\alpha} + 4\sum_{|k|\leq\frac{1}{h}} ||A_k||^2_{C_2}\frac{\pi^2 h^2}{16}k^2,$$

and we must only estimate

$$I_h = \sum_{|k|\leq n-1} h^2 k^2 ||A_k||^2_{C_2}.$$

Let

$$\gamma_k = \sum_{m=k}^\infty \{||A_m||^2_{C_2} + ||A_{-m}||^2_{C_2}\} \quad (k > 0),$$

such that

$$\gamma_k \leq C_1\frac{1}{k^{2\alpha}}.$$

Represent now I_h as

$$I_h = h^2\sum_{k=1}^{n-1} k^2(\gamma_k - \gamma_{k+1}).$$

Thus,

$$I_h \leq h^2\{\gamma_1 + \gamma_2(2^2 - 1) + \gamma_3(3^2 - 2^2) + \cdots + \gamma_{n-1}[(n-1)^2 - (n-2)^2]\}$$

$$\leq 2h^2\sum_{k=1}^{n-1} k\gamma_k \leq 2C_1 h^2\sum_{k=1}^{n-1}\frac{1}{k^{2\alpha-1}} \leq C'h^2(n-1)^{2-2\alpha} \leq C'h^{2\alpha},$$

where C' does not depend on h.

Hence,

$$\omega_2\left(\frac{\pi h}{2}\right) \leq h^\alpha\sqrt{8\pi C_1 + \frac{\pi^3}{2}C'},$$

and the proof is complete. $\qquad\qquad\qquad\qquad\qquad\qquad\qquad\qquad\qquad\qquad\square$

We note that

$$\min_{B_r} \int_0^{2\pi} ||A(x) - \sum_{r=-n+1}^{n-1} B_r e^{irx}||_{C_2}^2 dx = 2\pi \sum_{|k|\geq n} ||A_k||_{C_2}^2.$$

This formula represents the square of the best approximation in the 2-metric of the matrix A with respect to band type matrices of the order $< n$.

If we denote this last expression with $E_n^{(2)}[A]$, then the previous Theorem becomes:

If

$$\omega_2(\delta;\ A) \leq C\delta^\alpha \quad (0 < \alpha < 1),$$

then for every positive integer n

$$E_n^{(2)} \leq C\frac{1}{n^\alpha},$$

and conversely.

Notes

In this chapter we investigate some spaces of infinite matrices from the harmonic analysis point of view. Namely, in Section 3.2 we introduce the space of all infinite matrices corresponding to linear bounded operators such that they are approximable in the operator norm by finite type band matrices. For such matrices, which are extensions of periodical continuous functions on the torus, we develop a Fejer theory for this situation.

In Section 3.3 we introduce and study a space, which is an analogue of the classical space $L^1(\mathbb{T})$. For the matrices belonging to this space a version of Riemann-Lebesgue Lemma holds.

In Section 3.4 we briefly investigate the matrix version of functions of bounded variation and we prove an analogue of the well-known Jordan's Theorem about the uniform convergence of partial sums in the Fourier development of a continuous function.

Section 3.5 is dedicated to some problems concerning matriceal approximation and it extends the well-known Haar Theorem about the development of an essentially bounded function in a series of Haar functions. In order to give this theorem we introduce an interesting space of sequences, denoted by ms, and we study some matrices, which are extensions of usual scalars, from the point of view of Schur multipliers.

Finally, in Section 3.6 we investigate a class of matrices from the point of view of approximation with partial sums of its diagonals.

Chapter 4

Matrix versions of Hardy spaces

4.1 First properties of matriceal Hardy space

The results from this section were communicated to us by V. Lie.

We introduce a matrix version of the Hardy space, which will coincide with the classical one on the class of all Toeplitz matrices $\mathcal{T}$.

Let $A = (a_{jk})_{j \geq 1; \, k \geq 1}$ be an infinite matrix. We associate with A the matriceal periodical distribution (function) $\mathcal{L}_A(x, t)$ on $[0, 1] \times [0, 1]$ defined by

$$\mathcal{L}_A(x, t) := \sum_{k=1}^{\infty} \sum_{j=1}^{\infty} a_{kj} e^{2\pi ijx} e^{2\pi ikt}.$$

The above relation may be rewritten as

$$\mathcal{L}_A(x, t) = \sum_{k=1}^{\infty} \mathcal{L}_k^A(x) e^{2\pi ikt} = \sum_{j=1}^{\infty} \mathcal{C}_j^A(t) e^{2\pi ijx},$$

where $\mathcal{L}_k^A$ is the distribution (function) associated with row k whereas $\mathcal{C}_j^A$ is the distribution associated with column j.

Because we work only with upper triangular matrices it is convenient to define

$$\widetilde{\mathcal{L}}_k^A(x) := \mathcal{L}_k^A(x) e^{-2\pi ikx}.$$

Using these notations we have that

$$\mathcal{L}_A(x, t) = \sum_{k=1}^{\infty} \widetilde{\mathcal{L}}_k^A(x) e^{2\pi ik(x+t)}.$$

We remark that if A is an upper triangular Toeplitz matrix, then $\widetilde{\mathcal{L}}_k^A := \widetilde{\mathcal{L}}^A$ for all k, and $\mathcal{L}_A(x, t) = \widetilde{\mathcal{L}}^A(x) \sum_{k=1}^{\infty} e^{2\pi ik(x+t)}$.

The importance of matriceal distributions in what follows is stressed by the following equalities:

$$\mathcal{L}_{C*D}(t,s) = \int_0^1 \int_0^1 \mathcal{L}_C(t-\mu,v)\mathcal{L}_D(\mu,s-v)d\mu dv,$$

$$\mathcal{L}_{CD}(x,t) = \int_0^1 \mathcal{L}_D(x,s)\mathcal{L}_C(-s,t)ds.$$

For an infinite matrix A we denote by A' the matrix whose matriceal distribution is given by

$$\mathcal{L}_{A'}(x,t) := \sum_{k=1}^{\infty} \left(\widetilde{\mathcal{L}}_k^A\right)'(x)e^{2\pi ik(x+t)}.$$

We remark that the above definition coincides with the following implication:

$$\text{if } A = \sum_{k\in\mathbb{Z}} A_k, \text{ then } A' = \sum_{k\in\mathbb{Z}} kA_k,$$

where A_k is the kth diagonal of A.

Remark 4.1. *If we consider the distributions $\widetilde{\mathcal{L}}_k^A$ as acting on the torus, that is, if*

$$\widetilde{\mathcal{L}}_k^A(x) = \widetilde{\mathcal{L}}_k^A(e^{2\pi ix})$$

and if we work only with upper triangular matrices A, then we may regard $\widetilde{\mathcal{L}}_k^A$ as being a limit (in the space of distributions) at the border of an analytic distribution, that is

$$\lim_{r\to 1} \widetilde{\mathcal{L}}_k^A(re^{2\pi ix}) = \widetilde{\mathcal{L}}_k^A(e^{2\pi ix}).$$

In this way we arrive at the following notation:

Given an upper triangular matrix A we say that $A(r)$ $(0 < r < 1)$ is the analytic extension of A, if

$$\widetilde{\mathcal{L}}_{A(r)}(x,t) = \sum_{h=1}^{\infty} \widetilde{\mathcal{L}}_k^A(re^{2\pi ix})e^{2\pi ik(x+t)}.$$

Now we extend in the framework of matrices the definition of classical Lebesgue spaces L^p, for $1 \le p \le 2$.

Definition 4.2. *For $1 \le p \le 2$ we define the spaces $\mathcal{L}_r^p(\ell_2)$ as*

$$\mathcal{L}_r^p(\ell_2) := \{A \mid ||A||_{\mathcal{L}_r^p(\ell_2)} < \infty\},$$

where

$$||A||_{\mathcal{L}_r^p(\ell_2)} = \sup_{||x||_{\ell_2(\mathbb{N})}\le 1} \left(\int_0^1 \left(\sum_{k=1}^{\infty} |\mathcal{L}_k^A(s)|^2 |x_k|^2\right)^{p/2} ds\right)^{1/p}.$$

The above definition reflects the behaviour of the matrix A with respect to its rows; of course we may also define the spaces corresponding to the columns of A, namely $\mathcal{L}_c^p(\ell_2) := \{A \mid ||A||_{\mathcal{L}_c^p(\ell_2)} < \infty\}$, where

$$||A||_{\mathcal{L}_c^p(\ell_2)} = \sup_{||x||_{\ell_2(\mathbb{N})} \leq 1} \left(\int_0^1 \left(\sum_{k=1}^\infty |\mathcal{C}_k^A(s)|^2 |x_k|^2 \right)^{p/2} ds \right)^{1/p}.$$

Remark. Let A_0 be the main diagonal submatrix of a matrix $A \in \mathcal{L}^2(\ell_2)$. Since $||A_0||_{A \in \mathcal{L}^2(\ell_2)} = \sup_k |a_0^k|$, it follows that $\mathcal{L}^2(\ell_2)$ is not isomorphic to a Hilbert space. This remark will be of special interest in Chapter 6.

For simplicity in what follows we write $\mathcal{L}^p(\ell_2)$ instead of $\mathcal{L}_r^A(\ell_2)$.

Definition 4.3. *We define the matriceal Hardy space $H^p(\ell_2)$ of index p, $1 \leq p \leq 2$, in the following way:*

$$H^p(\ell_2) := \{A \mid A \text{ upper triangular} ; A \in \mathcal{L}^p(\ell_2)\}.$$

Here

$$||A||_{H^p(\ell_2)} := \sup_{||x||_{\ell_2(\mathbb{N})} \leq 1} \left(\int_0^1 \left(\sum_{k=1}^\infty |\mathcal{L}_k^A(s)|^2 |x_k|^2 \right)^{p/2} ds \right)^{1/p}.$$

An interesting property of the space $H^1(\ell_2)$ is the following Hardy-Littlewood type inequality:

Proposition 4.4. *Let $A \in H^1(\ell_2)$. Then we have that*

$$\sup_{||x|| \leq 1} \left\{ \int_0^1 (1-r) \left(\int_0^r \frac{\left(\int_0^1 \left(\sum_{j=1}^\infty |x_j|^2 |(\widetilde{\mathcal{L}}_j^A)'(se^{2\pi i\theta})|^2 \right)^{1/2} d\theta \right)}{1-s} ds \right)^2 dr \right\}^{1/2}$$

$$\leq ||A||_{H^1(\ell_2)}.$$

Proof. Let $x \in \ell_2(\mathbb{N})$ be given, so that $||x||_2 \leq 1$, and let $g \in L^2([0,1])$ be a fixed positive function such that $||g||_2 \leq 1$. Using Khintchine's inequality we get that, for all $s \in [0,1]$,

$$\int_0^1 (\sum_{j=1}^\infty |x_j|^2 |(\widetilde{\mathcal{L}}_j^A)'(se^{2\pi i\theta})|^2)^{1/2} d\theta \sim$$

$$\int_0^1 \int_0^1 |\sum_{j=1}^{\infty} x_j \epsilon_j(\omega)(\widetilde{\mathcal{L}}_j^A)'(se^{2\pi i\theta})|d\theta d\omega.$$

Here $\epsilon_j(\omega)$ stands for the jth Rademacher function.

By duality, it is enough to prove that

$$\underbrace{\int_0^1 \sqrt{1-r}g(r)\int_0^r \frac{\int_0^1 \int_0^1 |\sum_{j=1}^{\infty} x_j\epsilon_j(\omega)(\widetilde{\mathcal{L}}_j^A)'(se^{2\pi i\theta})|d\theta d\omega}{1-s}dsdr}_{I}$$

$$\leq ||A||_{H^1(\ell_2)}.$$

Applying Fubini's theorem we find that

$$I :=$$

$$\int_0^1 \left(\int_0^1 \sqrt{1-r}g(r)(\int_0^r \frac{\int_0^1 |\sum_{j=1}^{\infty} x_j\epsilon_j(\omega)(\widetilde{\mathcal{L}}_j^A)'(se^{2\pi i\theta})|d\theta}{1-s}ds)dr\right)d\omega \leq$$

$$\int_0^1 \left(\int_0^1 (1-r)\left(\int_0^r \frac{\int_0^1 |\sum_{j=1}^{\infty} x_j\epsilon_j(\omega)(\widetilde{\mathcal{L}}_j^A)'(se^{2\pi i\theta})|d\theta}{1-s}ds\right)^2 dr\right)^{1/2}d\omega.$$

We get, by using Cauchy-Schwarz inequality, that

$$\int_0^1 (1-r)\left(\int_0^r \frac{\int_0^1 |f'(se^{2\pi i\theta})|d\theta}{1-s}ds\right)^2 dr \leq$$

$$\int_0^1 (1-r)\left(\int_0^r \frac{ds}{(1-s)^2}\right)\left(\int_0^r \left(\int_0^1 |f'(se^{2\pi i\theta})|d\theta\right)^2 ds\right)dr \leq$$

$$\int_0^1 \left(\int_0^r \left(\int_0^1 |f'(se^{2\pi i\theta})|d\theta\right)^2 ds\right)dr \leq \text{(by Fubini's theorem)} \leq$$

$$\int_0^1 (1-r)\left(\int_0^1 |f'(re^{2\pi i\theta})|d\theta\right)^2 dr \leq \text{(by inequality (HL) on page 4)}$$

$$\leq C^2||f||_{H^1}^2.$$

Hence, denoting by $f(se^{2\pi i\theta}) = \sum_{j=1}^{\infty} x_j\epsilon_j(\omega)(\widetilde{\mathcal{L}}_j^A)(se^{2\pi i\theta})$, we get from the inequalities above that

$$I \leq C \int_0^1 \int_0^1 |\sum_{j=1}^{\infty} x_j\epsilon_j(\omega)(\widetilde{\mathcal{L}}_j^A)(e^{2\pi i\theta})|d\theta d\omega \sim \text{(by Fubini's theorem and}$$

$$\text{Khintchine's inequality)} \sim \int_0^1 \left(\sum_{j=1}^{\infty} |x_j|^2 |(\widetilde{\mathcal{L}}_j^A)(e^{2\pi i\theta})|^2\right)^{1/2} d\theta$$

$$\leq ||A||_{H^1(\ell_2)}||x||_2.$$

The proof is complete. $\square$

4.2 Hardy-Schatten spaces

In 1983 A. Shields proved an interesting inequality which holds in Schatten class C_1 (see [84]). This inequality is similar to the following well-known inequality of Hardy and Littlewood (see e.g. [96]):

$$\left(\sum_{n=0}^{\infty}(n+1)^{p-2}|a_n|^p\right)^{1/p} \leq C(p)\|f\|_{H^p}, \quad 0 < p \leq 2,$$

which holds for all functions $f(t) = \sum_{n\geq 0} a_n e^{int}$ belonging to the Hardy spaces $H^p(\mathbb{T})$. (See [23].)

The similarity between functions and infinite matrices was remarked for the first time by J. Arazy in [1] and A. Shields exploited this crucial idea further in [84].

It is natural to consider the Schatten class of order p as the similar notion of the Lebesgue space $L^p(\mathbb{T})$, $0 < p < \infty$ (see [84]). We note that the above analogy is not perfect since for $0 < p_1 < p_2 < \infty$, it follows that $L^{p_2}(\mathbb{T}) \subset L^{p_1}(\mathbb{T})$ contrarily to the known inclusion $C_{p_1} \subset C_{p_2}$.

Moreover, the analogue of the Riesz projection $P_0(f) := \sum_{n\geq 0} a_n e^{int}$, where f is a trigonometric polynomial, is the *triangular projection* $P_T(A) = (a'_{ij})_{i,j\geq 1}$, where

$$a'_{ij} = \begin{cases} a_{ij} & \text{if } 0 \leq j - i \leq k \\ 0 & \text{otherwise} \end{cases}, \quad A = (a_{ij})_{i,j\geq 1} \text{ and } a_{ij} = 0 \text{ if } |i-j| > k$$

for some fixed $k \in \mathbb{N}$.

Therefore, the analogue of the Hardy space $H^p(\mathbb{T})$, $0 < p < \infty$ is the space

$$T_p := \{A|\ A \text{ upper triangular matrix}; A \in C_p\},$$

with the norm $\|A\|_{T_p} = \|A\|_{C_p}$.

On the other hand, if $f(t) = a_n e^{int}$, then it follows that $\|f\|_{H^p} = |a_n|$, and, thus, the corresponding object to $|a_n|$ would be $\|A_n\|_{C_p} = \|(a_{i,i+n})_i\|_{\ell^p}$, where $A_n = (a_{i,i+n})_{i\geq 1}$, $n \in \mathbb{N}$.

Thus we may expect that the following inequalities hold:

$$\sum_{n=0}^{\infty}(n+1)^{p-2}\|A_n\|_{C_p}^p \leq K(p)\|A\|_{T_p}^p \tag{4.1}$$

for all $A \in T_p$, $1 \leq p \leq 2$, or, equivalently:

$$\sum_{n=0}^{\infty}(n+1)^{p-2}\left(\sum_{i=1}^{\infty}|a_{i,i+n}|^p\right) \leq K(p)\|A\|_{T_p}^p, \quad 1 \leq p \leq 2. \tag{4.2}$$

For $p = 1$ the inequality (4.2) holds with $K(1) = \pi$, that is

$$\sum_{n=0}^{\infty}(n+1)^{-1}\left(\sum_{i=1}^{\infty}|a_{i,i+n}|\right) \leq \pi||A||_{T_1}, \ A \in T_1$$

which was proved by Shields in 1983 (see [84]). We discuss this fact further in the next section. For another, more general, proof see [15]-Thm. 2.2-a) and also Section 4.4.

Now we briefly describe the main content of the section. First we consider the case $1 < p \leq 2$ and prove (only for $1 < p < 2$) a weaker result then (4.1). However, we conjecture that (4.1) holds also in general in the case $1 < p < 2$. Of course, (4.1) is easy to prove in the case $p = 2$ with $K(2) = 1$. Moreover, it is possible to state and prove a result dual to the first inequality.

In order to state and prove the result we have to recall some notations from the paper [53].

Let A be an upper triangular matrix and let $(A_k)_{k\geq 0}$ be the sequence of its diagonal matrices.

We denote by $T_p(\ell_R^2)$ the completion of the space of all finite sequences $(A_k)_{k=0}^n$ with respect to the norm

$$||(A_k)_{k=0}^n||_{T_p(\ell_R^2)} := ||(\sum_{k=0}^n A_k^* A_k)^{1/2}||_{C_p}, \text{ where } 1 \leq p \leq 2.$$

Here A_k^* is the adjoint matrix of A_k. It is clear that $T_p(\ell_R^2)$ is a space of upper triangular matrices.

Similarly, $T_p(\ell_C^2)$ is the completion of the space of all finite sequences $(A_k)_{k=0}^n$ with respect to the norm

$$||(A_k)_{k=0}^n||_{T_p(\ell_C^2)} := ||(\sum_{k=0}^n A_k A_k^*)^{1/2}||_{C_p}.$$

$T_p(\ell_C^2)$ is a space of upper triangular matrices too.

Now let us denote by $T_p(\ell_R^2) + T_p(\ell_C^2)$ the space of all upper triangular matrices A such that there exist $A' \in T_p(\ell_R^2)$ and $A'' \in T_p(\ell_C^2)$ with $A = A' + A''$. We introduce on this space the norm

$$||A||_{T_p(\ell_R^2)+T_p(\ell_C^2)} := \inf_{A_k=A_k'+A_k''}\{||(\sum_{k\geq 0}A_k'^* A_k')^{1/2}||_{C_p}+||(\sum_{k\geq 0}A_k''A_k''^*)^{1/2}||_{C_p}\},$$

$1 \leq p \leq 2$.

Let us remark that

$$||(\sum_{k=0}^{\infty}A_k^* A_k)^{1/2}||_{C_p} = \left[\sum_{j=1}^{\infty}\left(\sum_{i=1}^{j}|a_{ij}|^2\right)^{p/2}\right]^{1/p}, \text{ for } 1 \leq p \leq 2,$$

and, similarly, that

$$||(\sum_{k=0}^{\infty} A_k A_k^*)^{1/2}||_{C_p} = \left[\sum_{i=1}^{\infty}\left(\sum_{j=i}^{\infty}|a_{ij}|^2\right)^{p/2}\right]^{1/p},$$

so that

$$||A||_{T_p(\ell_R^2)+T_p(\ell_C^2)} = \inf_{A_k=A_k'+A_k'',k\geq 0}\left\{\left[\sum_{j=1}^{\infty}\left(\sum_{i=1}^{j}|a_{ij}'|^2\right)^{p/2}\right]^{1/p}\right. \tag{4.3}$$

$$\left. +\left[\sum_{i=1}^{\infty}\left(\sum_{j=i}^{\infty}|a_{ij}''|^2\right)^{p/2}\right]^{1/p}\right\}.$$

Moreover, the relation (I.13)-[53] implies that

$$\left(\sum_{k\geq 0}||A_k||_{C_p}^2\right)^{1/2} \leq ||A||_{T_p(\ell_R^2)+T_p(\ell_C^2)}. \tag{4.4}$$

We also note that

$$||A_n||_{T_p(\ell_R^2)+T_p(\ell_C^2)} = ||A_n||_{C_p} = \left(\sum_{i\geq 1}|a_{i,i+n}|^p\right)^{1/p}, \quad 1\leq p\leq 2. \tag{4.5}$$

Now we state and prove the following result:

Theorem 4.5. *Let $1 < p < 2$ and $A \in T_p(\ell_R^2) + T_p(\ell_C^2)$. Then there exists a positive constant $K(p)$ such that*

$$\sum_{n=0}^{\infty}(n+1)^{p-2}||A_n||_{C_p}^p \leq K(p)||A||_{T_p(\ell_R^2)+T_p(\ell_C^2)}. \tag{4.6}$$

Proof. We follow the idea of the proof in [23] pp. 95-97.

Let A be an upper triangular matrix such that $A = \sum_{k=0}^{n} A_k$, and let μ be the measure on $\mathbb{N}$ defined by

$$\mu(n) = \frac{1}{(n+1)^2}, \quad n = 0,1,2,\ldots$$

Let us consider $A(t) = \sum_{n=0}^{\infty} A_n e^{int} = A * E_t$, where $t \in \mathbb{T}$, $*$ means the Schur product of matrices and $E_t = (e_{kj})_{k,j\geq 1}$ is the Toeplitz matrix, where $e_{kj} = e^{i(j-k)t}$, for all $j, k \geq 1$. Therefore, $||A(t)||_{C_p} \leq ||A||_{T_p}$.

Now denote $(n+1)A_n$ by $\tilde{A}_n$ and fix $s > 0$. Then, $A(t) = \Phi_s(t) + \Psi_s(t)$, $t \in \mathbb{T}$, where

$$\Phi_s(t) = \begin{cases} A(t) & \text{if } ||A(t)||_{T_p} > s \\ 0 & \text{if } ||A(t)||_{T_p} \leq s \end{cases},$$

and $\Psi_s(t) = A(t) - \Phi_s(t)$, $t \in \mathbb{T}$.

It is clear that $[A(t)]_n = [\Phi_s(t)]_n + [\Psi_s(t)]_n$, $n = 0, 1, 2, \ldots$; and, therefore, $A_n * [E_t]_n = [\Phi_s]_n * [E_t]_n + [\Psi_s]_n * [E_t]_n$, where $[\Phi_s]_n = A_n$ and $[\Psi_s]_n = 0$ if $||A||_{T_p} = \sup_{t \in \mathbb{T}} ||A(t)||_{T_p} > s$. The same holds for Ψ_s.

Hence,

$$\sum_{n \geq 0} (n+1)^{p-2} ||A_n||_{C_p}^p = \sum_{n \geq 0} ||\tilde{A}_n||_{C_p}^p \mu(n) \qquad (4.7)$$

$$\leq 2^p \left(\sum_{n \geq 0} ||[\tilde{\Phi}_s]_n||_{C_p}^p \mu(n) + \sum_{n \geq 0} ||[\tilde{\Psi}_s]_n||_{C_p}^p \mu(n) \right).$$

Put now $\alpha(s) = \mu\{n; ||[\tilde{\Phi}_s]_n||_{C_p} > s\}$ and $\beta(s) = \mu\{n; ||[\tilde{\Psi}_s]_n||_{C_p} > s\} := \mu(E_s)$. Then

$$s^2 \beta(s) \leq \sum_{n \geq 0} ||[\Psi_s]_n||_{C_p}^2. \qquad (4.8)$$

We denote the set $\{n \geq 0; ||[\tilde{\Phi}_s]_n||_{C_p} > s\}$, by F_s.

Since $||A_n||_{C_p} \leq ||A||_{C_p}$, $n \in \mathbb{N}$, (see [34]) we have that

$$\mu(F_s) = \mu\{n; (n+1)||[\Phi_s]_n||_{C_p} > s\} \leq \mu\{n; (n+1)||\Phi_s||_{C_p} > s\} =$$

$$\sum_{\{n; ||\Phi_s||_{C_p} > \frac{s}{n+1}\}} \frac{1}{(n+1)^2} \leq 4 \sum_{\{n; ||\Phi_s||_{C_p} > \frac{s}{n+1}\}} \int_n^{n+1} \frac{1}{x^2} dx \leq \frac{4}{n_0 + 1},$$

where $n_0 = \min\{n; ||\Phi_s||_{C_p} > \frac{s}{n+1}\}$.

Thus,

$$\alpha(s) = \mu(F_s) \leq \frac{4||\Phi_s||_{C_p}}{s}, \quad s > 0. \qquad (4.9)$$

According to (4.9) we have that

$$\sum_{n=0}^{\infty} ||[\tilde{\Phi}_s]_n||_{C_p}^p \mu(n) = -\int_0^{\infty} s^p d\alpha(s) = p \int_0^{\infty} s^{p-1} \alpha(s) ds \leq$$

$$4p \int_0^{\infty} s^{p-1} \frac{||\Phi_s||_{C_p}}{s} ds \leq 4p \int_0^{||A||_{C_p}} ||A||_{C_p} s^{p-2} ds \leq \frac{4p}{p-1} ||A||_{T_p}^p,$$

for all $t \in \mathbb{T}$.

Moreover, by (4.8) and (4.4), it follows that

$$\sum_{n=0}^{\infty} ||[\tilde{\Psi}_s]_n||_{C_p}^p \mu(n) = p \int_0^{\infty} s^{p-1}\beta(s)ds \leq p \int_0^{\infty} s^{p-3} \sum_{n=0}^{\infty} ||[\Psi_s]_n||_{C_p}^2 ds =$$

$$p \int_{||A||_{C_p}}^{\infty} \sum_{n=0}^{\infty} ||A_n||_{C_p}^2 s^{p-3} ds \leq \frac{2p}{2-p} ||A||_{C_p}^{p-2} \sum_{n=0}^{\infty} ||A_n||_{C_p}^2 \leq$$

$$\frac{p}{2-p} ||A||_{T_p}^{p-2} ||A||_{T_p(\ell_R^2)+T_p(\ell_C^2)}^2.$$

By using (4.7) we get that

$$\sum_{n=0}^{\infty} (n+1)^{p-2} ||A_n||_{C_p}^p \leq K(p) \left[||A||_{T_p}^p + ||A||_{T_p}^{p-2} ||A||_{T_p(\ell_R^2)+T_p(\ell_C^2)}^2 \right]. \quad (4.10)$$

But in [38] it is proved that $||A||_{C_p} \leq ||A||_{T_p(\ell_R^2)+T_p(\ell_C^2)}$ and, consequently, we get the following inequality of Hardy-Littlewood type:

$$\sum_{n \geq 0} (n+1)^{p-2} ||A_n||_{C_p}^p \leq K(p) ||A||_{T_p(\ell_R^2)+T_p(\ell_C^2)}^p, \quad 1 < p < 2. \quad (4.11)$$

Since $T_p(\ell_R^2)+T_p(\ell_C^2)$ is the completion of the space of all finite sequences Theorem 4.5 is proved. $\qquad \square$

Remark 4.6. *Theorem 4.5 is weaker than the following strong version of the Hardy-Littlewood inequality: the inequality*

$$\sum_{n \geq 0} (n+1)^{p-2} ||A_n||_{C_p}^p \leq K(p) ||A||_{T_p}^p, \quad (4.12)$$

holds for all upper triangular matrices A.

Indeed in [38] it is proved that there is an upper triangular matrix $A \in C_p$ such that $A \notin T_p(\ell_R^2) + T_p(\ell_C^2)$, $1 < p < 2$.

Then it is natural to raise the following:

Question: *Does Theorem 4.5 hold for $1 < p < 2$ with (4.12) instead of (4.6)?*

Using the duality between $T_p(\ell_R^2) + T_p(\ell_C^2)$ and $T_p(\ell_R^2) \cap T_p(\ell_C^2)$ (see [51]) given by

$$< A, B > = \sum_{k=0}^{\infty} tr\, (A_k B_k^*),$$

where $A = \sum_{k \geq 0} A_k$, $B = \sum_{k \geq 0} B_k$, we have the following:

Theorem 4.7. *Let* $2 \leq q < \infty$ *and* $A = \sum_{k \geq 0} A_k$ *such that* $\sum_{n \geq 0}(n + 1)^{q-2}\|A_n\|^q_{C_q} < \infty.$
Then $A \in T_q(\ell^2_R) \cap T_q(\ell^2_C)$ *and*

$$\|A\|_{T_q(\ell^2_R) \cap T_q(\ell^2_C)} := \max\left((\sum_{i=1}^{\infty}(\sum_{j=i}^{\infty}|a_{i,j}|^2)^{q/2})^{1/q}, (\sum_{j=1}^{\infty}(\sum_{i=1}^{j}|a_{i,j}|^2)^{q/2})^{1/q} \right)$$

$$\leq C(q)\left(\sum_{n \geq 0}(n+1)^{q-2}\|A_n\|^q_{C_q} \right)^{1/q}.$$

Proof.   Let $p = q/(q-1)$ and $G = \sum_{k=0}^{n} G_k$ be a finite type band matrix with $\|G\|_{T_p(\ell^2_R) + T_p(\ell^2_C)} \leq 1$.
   Let $S_n(A) = \sum_{k=0}^{n} A_k$. Then

$$| < G, S_n(A) > | = |\sum_{k=0}^{n} tr\,(G_k A_k^*)| \leq \sum_{k=0}^{n} |\sum_{i=1}^{\infty} g_{i,k+i}\overline{a_{i,k+i}}| \leq \text{(by Hölder's}$$

$$\text{inequality)} \leq \sum_{k=0}^{n}(\sum_i |g_{i,k+i}|^p)^{1/p}(\sum_i |a_{i,k+i}|^q)^{1/q} = \sum_{k=0}^{n} \|G_k\|_{C_p}\|A_k\|_{C_q}$$

$$\leq \text{(again by Hölder's inequality)} \leq (\sum_{k=0}^{n} \|G_k\|^p_{C_p}(k+1)^{p-2})^{1/p}$$

$$(\sum_{k=0}^{n} \|A_k\|^q_{C_q}(k+1)^{q-2})^{1/q} \leq \text{(by Theorem 4.5)}$$

$$\leq C(p)\|G\|_{T_p(\ell^2_R) + T_p(\ell^2_C)}\left(\sum_{k=0}^{n} \|A_k\|^q_{C_q}(k+1)^{q-2} \right)^{1/q}$$

$$\leq C(p)\left(\sum_{k=0}^{n} \|A_k\|^q_{C_q}(k+1)^{q-2} \right)^{1/q}.$$

Hence,

$$\|S_n(A)\|_{T_q(\ell^2_R) \cap T_q(\ell^2_C)} = \sup_{\|G\|_{T_p(\ell^2_R) + T_p(\ell^2_C)} \leq 1} | < G, S_n(A) > |$$

$$\leq C(p)\left(\sum_{k=0}^{n} \|A_k\|^q_{C_q}(k+1)^{q-2} \right)^{1/q},$$

for all n and the proof is complete. $\square$

Now we discuss an inequality of Hausdorff-Young type.

Theorem 4.8. *(Hausdorff-Young's inequality) For $1 \leq p \leq \infty$, let q be the conjugate index, with $\frac{1}{p} + \frac{1}{q} = 1$.*

(i) *If $1 \leq p \leq 2$, then $A \in T_p$ implies that $\left(\sum_{n=0}^{\infty} ||A_n||_{T_p}^q \right)^{1/q} \leq ||A||_{T_p}$.*

(ii) *If $2 \leq p \leq \infty$, then $\{||A_n||_{T_p}\} \in \ell_q$ implies that*

$$||A||_{T_p} \leq \left(\sum_{n=0}^{\infty} ||A_n||_{T_p}^q \right)^{1/q} .$$

Proof. In case (ii), for $p = q = 2$, if $\left(\sum_{n=0}^{\infty} \sum_{l=1}^{\infty} |a_{l,n+l}|^2 \right)^{1/2} < \infty$, then clearly $A = (a_{ij}) \in T_2$ and $||A||_{T_2} \leq \left(\sum_{n=0}^{\infty} \sum_{l=1}^{\infty} |a_{l,n+l}|^2 \right)^{1/2}$, in other words the map $T(\{A_n\}_{n \geq 0}) = \sum_{n \geq 0} A_n$ has norm less than 1 from the space $\ell_2(\ell_2)$ into T_2. Here $\ell_2(\ell_2)$ means the space of all matrices $(a_{ij})_{i,j}$ such that $\sum_{n=1}^{\infty} \sum_{i=1}^{\infty} |a_{i,i+n}|^2 < \infty$.

If $p = \infty$ it follows that $q = 1$ and the map T has the norm less than 1 from $\ell_1(\ell_\infty)$ into T_∞, since, clearly, $||A||_{T_\infty} \leq \sum_{n=0}^{\infty} ||A_n||_{T_\infty}$.

Using complex interpolation (see [19]), we get the conclusion.

Case (i) follows by duality, since $\ell_p(\ell_q)^* = \ell_q(\ell_p)$, and $(T_p)^* = T_q$. $\quad\square$

4.3 An analogue of the Hardy inequality in T_1

In this section we present an important inequality due to A. Shields [84]. In fact, the paper of Shields containing this inequality was the starting point of the material described in the present book.

Let us state the analogue of the inequality of Hardy, Littlewood and Fejér. In the next section we give also a more general useful inequality, which will be proved using different methods.

Theorem 4.9. *Let $M \in C_1$ have the upper triangular form with respect to the orthonormal basis $\{e_n\}$ $(n = 1, 2, \ldots)$ of ℓ_2. Then*

$$\sum_{k=1}^{\infty} \sum_{j=1}^{k} \frac{|M(j,k)|}{1 + k - j} \leq \pi ||M||_{T_1},$$

with equality only when $M = 0$.

It is easy to observe that another form of the above inequality is as follows:

$$\sum_{k=0}^{\infty} \frac{1}{k+1} ||M_k||_{T_1} \leq \pi ||M||_{T_1}.$$

This inequality is similar with the classical inequality of Hardy [23].

In order to prove Theorem 4.9 we require three lemmas. Throughout our discussion the orthonormal basis $\{e_n\}$, $n = 1, 2, \ldots$ will be fixed; upper triangularity will always be with respect to this basis.

Lemma 4.10. *Let R denote either the space $B(\ell_2)$, with the weak operator topology, or any of the Banach spaces C_p ($1 < p < \infty$) with its weak topology. If $\{A^n\}$, $\{B^n\} \subseteq R$, with $A^n \to A$ and $B^n \to B$ weakly, and if each B^n has the upper triangular form, then $A^n B^n \to AB$ weakly.*

Proof.   We write as usual in this book operators as matrices. One can easily verify that if $\{A^n\} \subset R$, then $A^n \to A$ weakly if and only if $\{||A^n||_R\}$ is a bounded sequence, and $A^n(i, j) \to A(i, j)$ for all i, j. Thus to complete the proof we must show that

a) $||A^n B^n||_R$ are bounded, and

b) $d_n(i, j) \to d(i, j)$ for all i, j.

Here we let d_n and d denote the matrix entries of $A^n B^n$ and of AB.

Concerning the first point we recall that in C_p we have that

$$||A^n B^n||_{T_p} \leq ||A^n B^n||_{C_{p/2}} \leq ||A^n||_{C_p} ||B^n||_{C_p}$$

(see [34], Chap. III, (7.4) and (7.5)). Thus $\{||A^n B^n||_R\}$ is a bounded sequence.

For the second point we note that

$$d_n(i, j) = \sum_{k=1}^{j} A^n(i, k) B^n(k, j)$$

since B^n is upper triangular. A similar equation holds for $d(i, j)$, and, thus, for each fixed choice of i, j, it yields that $d_n(i, j) \to d(i, j)$. The proof is complete. $\square$

Lemma 4.11. *Let $P \in C_1$ with $||P||_{T_1} = 1$, be a positive semi-definite operator. Then there exists $B \in C_2$ with $||B||_{C_2} = 1$ such that B has the upper triangular form and $P = B^*B$.*

Proof. Let us denote by E_n the subspace $span\{e_1, ..., e_n\}$, for all $n \in \mathbb{N}$. We first prove the lemma under the additional assumption that P is one-to-one on each of the subspaces E_n $(n \geq 1)$. Then $P^{1/2}$ is also one-to-one on each of these spaces. Let $F_n = P^{1/2}E_n$. Then $F_1 \subset F_2 \subset ...$, and $dim\ F_n$ for all n. Hence, there is an orthonormal set $\{f_k\}$ such that $F_n = span\{f_1, ..., f_n\}$. Define an operator V by: $V f_n = e_n$ $(n \geq 1)$, and $V = 0$ on the orthogonal complement of the span of $\{f_n\}$. Then V is a partial isometry. Let $B = VP^{1/2}$. Then $BE_n \subset E_n$ and, thus, B has the upper triangular form. Moreover, $B^*B = P^{1/2}V^*VP^{1/2} = P$, since V^*V is the projection onto the span of $\{f_n\}$, which contains the range of $P^{1/2}$. Finally, $||B||_{T_2} \leq ||V||\ ||P^{1/2}||_{T_2} = 1$, and, therefore, $1 = ||P||_{T_1} \leq ||B^*||_{T_2}||T||_{T_2} \leq 1$, which completes the proof in this case.

Now suppose that P is not one-to-one. Let S be a fixed positive operator from C_1 with trivial kernel. Let $P^n = (P + n^{-1}S)d_n$, where $d_n = ||P + n^{-1}S||_{T_1}^{-1}$. Then P^n has norm one, and $P^n \to P$, in C_1. In view of the result proved above, there is a sequence $\{B^n\}$ of operators in C_2, having the upper triangular form, with $P^n = (B^n)^*B^n$, $||B^n||_{T_2} = 1$, for all n. By passing to a subsequence we may assume that $\{B^n\}$ is weakly convergent in $C_2 : B^n \to B$ for some B in the unit ball of C_2. The limit operator B must have the upper triangular form and, by Lemma 4.10, we have that $B^*B = P$. From this we have that $||B||_{T_2} \geq 1$ and, hence, the norm must be equal to unity. The proof is complete. $\square$

Lemma 4.12. *Let $M \in C_1$ have the upper triangular form with $||M||_{T_1} = 1$. Then there exist upper triangular operators $A, B \in C_2$ with $M = AB$ and $||A||_{T_2} = ||B||_{T_2} = 1$.*

Proof. We first prove the lemma with the additional assumption that M is one-to-one on each of the spaces E_n; this is equivalent to requiring that all diagonal matrix entries are different from 0; $< Me_j, e_j > \neq 0$ for all j.

Let $M = UP$ be the polar decomposition of M. Then $P = (M^*M)^{1/2}$ is a positive operator of norm one in C_1 and U maps the range of P isometrically onto the range of M. Since $||Pf|| = ||Mf||$ for all f we see that P has the same kernel as M. Therefore, P is one-to-one on each of the spaces E_n. By Lemma 3.3, $P = B^*B$, where B is an upper triangular operator of norm one in C_2. We see that B must be one-to-one on each of the spaces E_n. Now let $A = UB^*$. Then A is in the unit ball of C_2, and $AB = M$. From this we see that $||A||_{C_2} = 1$. To show that A has the upper triangular form we must show that it maps each space E_n into itself. Since B is one-to-one on E_n and E_n is finite-dimensional we have that $E_n = BE_n$. Hence,

$$AE_n = ABE_n = ME_n = E_n.$$

Now suppose that M is not one-to-one on each of the spaces E_n, that is, some diagonal matrix entries are 0. Let S be a matrix from C_1, with non-zero diagonal entries precisely in those places where M has a zero. Let $M^n = (M + n^{-1}S)d_n$, where $d_n = ||M + n^{-1}S||_{C_1}^{-1}$. Then M^n satisfies all the conditions of the lemma and, in addition, is one-to-one on each of the spaces E_n. By what was proved above there are upper triangular operators A^n, B^n in the unit ball of C_2, $Ball(C_2)$, with $M^n = A^n B^n$. By passing to a subsequence we may assume that the sequences $\{A^n\}$ and $\{B^n\}$ are weakly convergent in $C_2 : A^n \to A$, $B^n \to B$, where $A, B \in Ball(C_2)$. By Lemma 4.10 we have that $A^n B^n \to AB$, and so $M = AB$. This completes the proof since weak convergence preserves the upper triangular form.   □

Proof of Theorem 4.9. Without loss of generality we may assume that $||M||_{T_1} = 1$. Then by Lemma 4.12 there are upper triangular operators A, B of norm one in C_2 such that $M = AB$. Let m_{ij}, a_{ij}, b_{ij} denote the matrix entries of M, A, B, respectively. The following summations are written with each variable going from 1 to ∞. Because of the upper triangularity, however, the terms are equal to zero if $j > k$, or if $j > r$, or if $r > k$. Thus, we really have that $1 \leq j \leq r \leq k < \infty$. We use the boundedness of the second Hilbert matrix, that is the matrix with entries $(n - m)^{-1}$ when $n \neq m$, and 0 when $n = m$ $(n, m = 1, 2, \dots)$; then we use the Cauchy-Buniakovsky-Schwarz inequality and obvious estimates to obtain that

$$\sum_{j,k} \frac{|m_{jk}|}{1 + k - j} \leq \sum_{j,k} \sum_{r} \frac{|a_{jr} b_{rk}|}{1 + k - j} = \sum_{r} \sum_{j,k} \frac{|a_{jr} b_{rk}|}{1 + k - j}$$

$$\leq \pi \sum_{r} \left(\sum_{j} |a_{jr}|^2 \right)^{1/2} \left(\sum_{k} |b_{rk}|^2 \right)^{1/2}$$

$$\leq \pi \left(\sum_{r} \sum_{j} |a_{jr}|^2 \right)^{1/2} \left(\sum_{r} \sum_{k} |b_{rk}|^2 \right)^{1/2} = \pi.$$

We have strict inequality because the bound π for the second Hilbert matrix is not attended. The proof is complete.   □

4.4 The Hardy inequality for matrix-valued analytic functions

Here we present another proof of the previous inequality of Shields. In fact we present a more general inequality for vector-valued analytic functions emphasizing the special case of matrix Banach spaces.

All the results of this section are due to O. Blasco and A. Pelczynski [15].

We recall that if $f = \sum_{j \geq 0} a_j e^{ijt}$ is an analytic trigonometric polynomial, then

$$\sum_{j \geq 0} |a_j|(j+1)^{-1} \leq C_1 \int_{-\pi}^{\pi} |f(t)| \, dt,$$

$$\left(\sum_{k \geq 0} |a_{2^k}|^2 \right)^{1/2} \leq C_2 \int_{-\pi}^{\pi} |f(t)| \, dt,$$

where C_1 and C_2 are numerical constants independent of f (cf. [23]). The first fact is called the Hardy inequality; the second is a particular case of a theorem of Paley, where (2^k) is replaced by any sequence (n_k) of positive integers with $\inf_k n_{k+1}/n_k > 1$. It is also known that both of these inequalities are false if analytic trigonometric polynomials are replaced by arbitrary trigonometric polynomials.

In what follows we are interested in finding under which additional conditions on a Banach space X the inequalities remain true if the Fourier coefficients a_j's are elements of X and absolute values are everywhere replaced by norms. We remark that it is known that in that general setting for arbitrary Banach spaces the inequalities are false (see for instance [15]).

It appears that the validity of X-valued versions of these inequalities depends on geometric properties of X. We are specially interested in the case when X is some matrix space, for instance if $X = C_1$, the Banach space of all trace class matrices.

The main idea of the proofs is to use vector-valued Hardy spaces H^p_X and to consider and use some operators induced by bounded multipliers from H^1 into ℓ_1.

4.4.1 *Vector-valued Hardy spaces H^p_X*

All Banach spaces are considered to be taken over the complex number field $\mathbb{C}$. Given a Banach space X and $p \in [1, \infty)$ (respectively $p = \infty$) we denote by L_X^p the space of all X-valued 2π-periodic functions on the real line $\mathbb{R}$, which are Bochner absolutely integrable in the pth power (respectively essentially bounded) under the norm

$$\|f\|_p = \left[(2\pi)^{-1}\int_{-\pi}^{\pi}\|f(t)\|^p dt\right]^{1/p} \quad \text{for } 1 \le p < \infty$$

(respectively $\|f\|_\infty = \text{ess sup}_{t \in \mathbb{R}}\|f(t)\|$).

Given $f \in L_X^1$ and an integer j, the jth Fourier coefficient of f is defined by

$$\widehat{f}(j) = (2\pi)^{-1}\int_{-\pi}^{\pi}e^{-ijt}f(t)dt.$$

If for some nonnegative integer n, $\widehat{f}(j) = 0$ for $|j| > n$, then f is called an X-valued trigonometric polynomial of degree $\le n$; if, moreover, $\widehat{f}(j) = 0$ for $j < 0$, then f is called an X-valued analytic trigonometric polynomial.

Given $p \in [1, \infty)$ the Hardy space H_X^p is defined to be the closure of all X-valued analytic trigonometric polynomials under the norm $\|\cdot\|_p$; or in other words

$$H_X^p = \{f \in L_X^p : \widehat{f}(j) = 0 \text{ for } j < 0\}.$$

4.4.2 $(H^p - \ell_q)$-multipliers and induced operators for vector-valued functions

Let $m = (m_j)_{j \ge 0}$ be a complex sequence and let X be a Banach space. Define the operator m_X from X-valued analytic trigonometric polynomials into the eventually zero X-valued sequences by $m_X(f) = (m_j\widehat{f}(j))_{j \ge 0}$. We call m_X the operator induced by the multiplier m. The operator m_X is said to be (p, q)-bounded provided that there exists a constant $K = K(m, X)$ such that, for every X-valued analytic trigonometric polynomial f,

$$\left(\sum_{j=0}^{\infty}\|m_j\widehat{f}(j)\|^q\right)^{1/q} \le K\|f\|_p$$

If m_X is (p, q)-bounded, then it can be uniquely extended to an operator (also denoted by m_X) from H_X^p into the Banach space $(\ell_q)_X$, where

$$(\ell_q)_X = \{(x_j) \subset X : \|(x_j)\|_q = \left(\sum\|x_j\|^q\right)^{1/q} < \infty\}.$$

We call m an $(H^p - \ell_q)$-multiplier if, for $X = \mathbb{C}$, $m_{\mathbb{C}}$ is (p, q)-bounded.

Definition 4.13. *A Banach space X is of $(H^1 - \ell_1)$-Fourier type provided that, for every $(H^1 - \ell_1)$-multiplier m, the induced multiplier m_X is $(1, 1)$-bounded.*

Recall the following beautiful description of $(H^1 - \ell_1)$-multipliers given by Ch. Fefferman in an unpublished manuscript:

Theorem 4.14. *A scalar sequence* $m = (m_j)_{j \geq 0}$ *is an* $(H^1 - \ell_1)$ *multiplier if and only if*

$$\rho(m) = \left(|m_0|^2 + |m_1|^2 + \sup_{s \geq 1} \sum_{k=1}^{\infty} \left(\sum_{j=ks+1}^{(k+1)s} |m_j| \right)^2 \right)^{1/2} < \infty.$$

Following the lines of the paper [88] we give a proof of the above theorem.

Let us denote by Λ the lattice of all integers of $\mathbb{R}$ and by Q_α^ϵ the interval $\{x \in \mathbb{R} : \epsilon\alpha - \epsilon/2 \leq x < \epsilon\alpha + \epsilon/2\}$, where $\alpha \in \Lambda$ and $\epsilon > 0$.

Now we state the following theorem belonging to Sledd and Stegenga [88]:

Theorem 4.15. *Let* μ *be a positive Borel measure on* $\mathbb{R} \setminus \{0\}$. *Then*

$$\sup_{||f||_{H^1(\mathbb{R})} \leq 1} \int |\widehat{f}| d\mu < \infty \tag{4.13}$$

if and only if

$$\sup_{\epsilon > 0} \left(\sum \mu(Q_\alpha^\epsilon)^2 \right)^{1/2} < \infty. \tag{4.14}$$

Moreover, the corresponding suprema are equivalent.

Corollary 4.16. *Let* $\{m_\alpha\}_{\alpha \in \Lambda}$ *be nonnegative numbers and define a measure on* $\mathbb{R} \setminus \{0\}$ *by* $\mu = \sum_{\alpha \neq 0} m_\alpha \delta_\alpha$, *where* δ_α *is the point mass at* $x = \alpha$. *Then*

$$\sup_{||f||_{H^1(\mathbb{T})} \leq 1} \sum_{\alpha \neq 0} |\widehat{f}(\alpha)| m_\alpha < \infty \tag{4.15}$$

if and only if μ *satisfies condition* (4.14).

It is easy to see that Corollary 4.16 is nothing else than Theorem 4.14. Now we proceed to the proof of Theorem 4.15.

Proof. We recall that an atom $a(x)$ corresponding to an interval Q is a measurable function supported on Q which has zero mean and is bounded by $|Q|^{-1}$ ($| \cdot |$ meaning the Lebesgue measure). By a fundamental result of R. Coifman [21] we may take as a definition of a function f of $H^1(\mathbb{R})$ the equality $f = \sum_i \lambda_i a_i$, where $\sum_i |\lambda_i| < \infty$ and $||f||_{H^1(\mathbb{R})} = \inf\{\sum_i |\lambda_i|\}$, for all $\{\lambda_i\}$ as before.

Thus, the sufficiency of condition (4.14) follows if there is a $c < \infty$ with

$$\int |\widehat{a}| d\mu \le c \qquad (4.16)$$

for all atoms a.

Part of the proof of (4.16) is straightforward. If a is an atom corresponding to an interval of length δ, then it is easy to see that $|\widehat{a}(y)| \le c|y|\delta$ for $y \in Q_0^\epsilon$, where $\epsilon = \delta^{-1}$. Here c is a constant nondepending of a. Now it is clear that (4.14) implies

$$\epsilon^{-1} \int_{Q_0^\epsilon} |x| d\mu(x) \le c$$

and, hence, (4.16) will follow from

$$\int_{\mathbb{R} \backslash Q_0^\epsilon} |\widehat{a}| d\mu \le c \qquad (4.17)$$

where ϵ is related to a as above. This result is now easily seen to be a consequence of condition (4.14) and the following theorem

Theorem 4.17. *There is a constant $c < \infty$ such that if $a(x)$ is an atom corresponding to an interval with the length 2δ and $\epsilon = \delta^{-1}$, then*

$$\sum_\alpha \sup_{Q_\alpha^\epsilon} |\widehat{a}|^2 \le c.$$

Proof. It suffices to assume that a is smooth and supported in the interval $[-\delta/2, \delta/2]$.

Fix an interval I of length ϵ and assume that f is continuously differentiable on I. It is elementary to see that $\sup_I |f - b| \le \int_I |f'|$, where b is the average $|I|^{-1} \int_I f$. Hence,

$$\sup_I |f|^2 \le 2 \left[\frac{1}{\epsilon} \int_I |f|^2 + \epsilon \int_I |f'|^2 \right].$$

Normalizing the Fourier transform so that $\|f\|_2 = \|\widehat{f}\|_2$, we obtain that

$$\sum_\alpha \sup_{Q_\alpha^\epsilon} |\widehat{a}|^2 \le 2 \left[\frac{1}{\epsilon} \int_{-\infty}^{\infty} |\widehat{a}|^2 + \epsilon \int_{-\infty}^{\infty} |\widehat{a}'|^2 \right]$$

$$= 2 \left[\frac{1}{\epsilon} \int_{-\delta/2}^{\delta/2} |a|^2 + \epsilon \int_{-\delta/2}^{\delta/2} |2\pi i x a|^2 dx \right]$$

from which Theorem 4.17 follows. $\square$

The following Lemma is useful for our purposes.

Lemma 4.18. *Let* $g \in L^2(\mathbb{R})$ *and assume that* $\widehat{g} = 0$ *on* $|y| \leq 1$. *If* $f = g\widehat{\chi}_{[-1,1]}$, *then* $f \in H^1$ *and* $||f||_{H^1} \leq c||g||_2$. *(Here* $\chi_{[-1,1]}$ *is the characteristic function for the unit interval centered at the origin.)*

Proof. Assume that $\widehat{g}$ is a C^∞-function with compact support in $|y| > 1$. Then f is the convolution $\widehat{g} * \chi_{[-1,1]}$ and, hence, is a rapidly decreasing function, which vanishes in a neighborhood of the origin. Thus f is in H^1. If $u \in BMO(\mathbb{R})$ and b is its average over $[-1,1]$, then, by the Schwarz inequality,

$$|\int fu| = |\int f(u-b)| \leq c||g||_2 \left\{ \int \frac{|u-b|^2}{1+|x|^2} dx \right\}^{1/2} \leq c||g||_2 ||u||_{BMO}.$$

The first inequality is a well-known estimate for $\widehat{\chi}_{[-1,1]}$ and the second a slight extension of inequality (1.2) in [32]. Now use the duality. □

The proof of Theorem 4.15 is complete once we establish the necessity of the condition (4.14). However, if (4.13) holds with supremum A, then from Lemma 4.18 we deduce that

$$\int [\mu(y + [-1,1])]^2 dy \leq cA^2. \tag{4.18}$$

Hence, there is an $M < \infty$, $\delta > 0$, for which $\sum_{|\alpha| \geq M} \mu(Q_\alpha^\delta) \leq cA^2$, where c is an absolute constant. But then a dilation argument gives this inequality for all $\delta > 0$ and (4.14) now follows in an elementary way. Thus, also the proof of Theorem 4.15 is complete. □

Proof of Corollary 4.16. The space $H_0^1(\mathbb{T})$ is the subspace of $H^1(\mathbb{T})$ consisting of functions with zero mean. Given $f \in H^1(\mathbb{R})$, we define

$$Pf(x) := \sum_{\alpha \in \Lambda} f(x + \alpha).$$

Since $f \in L^1(\mathbb{R})$ we have that $Pf \in L^1(\mathbb{T})$ and, by the Poisson summation formula (see [90]), it follows that $\widehat{f}(\alpha) = (Pf)\widehat{}(\alpha)$ for $\alpha \in \Lambda$.

The proof of the corollary is an immediate consequence of the following theorem:

Theorem 4.19. *It yields that* $P(H^1(\mathbb{R})) = H_0^1(\mathbb{T})$.

Proof. Let ϕ be a nonnegative rapidly decreasing function for which $\widehat{\phi}$ has support contained in the open interval $(-1,1)$ and $\widehat{\phi}(0) = 1$. Put $\varphi_\epsilon(x) = \epsilon\varphi(\epsilon x)$ for $0 < \epsilon < 1$. For a polynomial $F(x) = \sum a_\alpha e^{2\pi i \alpha x}$ let $f_\epsilon = F * \varphi_\epsilon$.

Claim. $\lim_{\epsilon \to 0} \|f_\epsilon\|_{H^1(\mathbb{R})} \leq \|F\|_{H^1(\mathbb{T})}$.

We start with the easily derived fact that

$$\lim_{\epsilon \to 0} \int_{\mathbb{R}} g|\varphi_\epsilon| = \int_{\mathbb{T}} g\,dx \qquad (4.19)$$

for all continuous functions g on $\mathbb{T}$. Observe that $\|\varphi_\epsilon\|_1 = 1$.

Let S, R denote the Riesz projections on $\mathbb{T}$, $\mathbb{R}$. Then, by (4.19), we obtain that

$$\limsup_{\epsilon \to 0} [\|f_\epsilon\|_1 + \|Rf_\epsilon\|_1] \leq \|F\|_{H^1(\mathbb{T})} + \limsup_{\epsilon \to 0} \|Rf_\epsilon - (SF)(\varphi_\epsilon)\|_1$$

so that we must show that the second term on the right hand side is zero. Since F is a polynomial it suffices to fix $\alpha \in \Lambda$ with $\alpha \neq 0$, put

$$h_\epsilon(y - \alpha) = (y/|y| - \alpha|\alpha|)\widehat{\varphi}_\epsilon(y - \alpha)$$

and show that $\lim_{\epsilon \to 0} \|\widehat{h}_\epsilon\|_1 = 0$.

Now $\widehat{\varphi}_\epsilon$ is supported in $[-\epsilon, \epsilon]$. Thus, we may assume that $h_\epsilon(y) = m(y)\widehat{\varphi}_\epsilon(y)$, where m is smooth, all derivatives up to order 2 are bounded by a constant, and $|m(y)| \leq c|y|$. The conditions on m imply that $\|Dh_\epsilon\|_1 \leq c$, where $D = d^2/dy^2$. Hence, $|\widehat{h}_\epsilon(x)| \leq c|x|^{-2}$. Clearly, $\lim_{\epsilon \to 0} \|h_\epsilon\|_1 = 0$ so that $\lim_{\epsilon \to 0} \|\widehat{h}_\epsilon\|_\infty = 0$ and, thus, the above estimate implies that $\lim_{\epsilon \to 0} \|\widehat{h}_\epsilon\|_1 = 0$. This proves the claim.

To complete the proof we fix F in $H_0^1(\mathbb{T})$ and note that there are polynomials $F_n \in H_0^1(\mathbb{T})$ with $\sum \|F_n\|_{H^1} < \infty$ and $F = \sum F_n$. Using the above we find that $f_n \in H^1(\mathbb{R})$ with $\sum \|f_n\|_{H^1(\mathbb{R})} < \infty$ and $Pf_n = F_n$. Thus, $Pf = F$ where $f = \sum f_n$ is a function in $H^1(\mathbb{R})$. The proof is complete. $\square$

By summing up we note that also the proof of Theorem 4.14 is complete.
$\square$

In the sequel FM means the Banach space of all scalar sequences satisfying the above relation equipped with the norm $\rho(\cdot)$.

Now we present a dual description of $(H^1 - \ell_1)$ Fourier type spaces.

Proposition 4.20. *For every Banach space X the following statements are equivalent:*

(i) X is an $(H^1 - \ell_1)$ Fourier type space;

(ii) there is $C > 0$ such that, for every $m \in FM$ and $f \in H_X^1$,

$$\sum_{j \geq 0} \|m_j\widehat{f}(j)\| \leq C\rho(m)\|f\|_1;$$

(iii) there is $C > 0$ such that for every eventually zero sequence $(x_j^*)_{j \geq 0}$ of elements of X^* there is an X^*-valued trigonometric polynomial g^* such that

$$\widehat{g^*} = x_j^* \text{ for } j \geq 0; \; \|g^*\|_\infty \leq C\rho((\|x_j^*\|)_{j \geq 0}).$$

Proof. $(i) \Rightarrow (ii)$. Put

$$\rho_X(m) = \sup\{\sum_{j \geq 0} ||m_j \widehat{f}(j)||; f \in H_X^1; ||f||_1 = 1\}.$$

Using the Baire category argument we get that $\rho_X(\cdot)$ is a bounded norm on FM.

$(ii) \Rightarrow (iii)$. Let $x_j^* = 0$ for $j \geq N$. We define on H_X^1 the linear functional ϕ_0^* by $\phi_0^*(f) = \sum_{j=0}^N x_j^*(\widehat{f}_j)$ for $f \in H_X^1$. It follows from (ii) that $||\phi_0^*|| \leq C\rho((||x_j^*||)_{j \geq 0})$. Let ϕ^* be a norm-preserving extension of ϕ_0^* onto L_X^1. Let V be the Nth de la Vallè Poussin kernel, i.e., $\widehat{V}(j) = 1$ for $|j| \leq N$, $\widehat{V}(j) = 0$ for $|j| \geq 2N$ and $\widehat{V}(j)$ linear for $-2N \leq j \leq -N$ and for $N \leq j \leq 2N$. It is well known that $||V||_1 \leq 2$. Define, for $j = 0, \pm 1, \pm 2, \ldots, y_j^* \in X^*$ by

$$y_j^*(x) = \widehat{V}(j)\phi^*(xe_j) \quad \text{for } x \in X,$$

where $e_j(t) = e^{ijt}$. Put $g^*(t) = \sum_{|j| \leq 2N} y_j^* e^{ijt}$. Then (denoting by $a * b$ the convolution of the functions a and b)

$$< f, g^* > = \phi^*(V * f) \quad \text{for } f \in L_X^1.$$

Thus, using the inequality $||V||_1 \leq 2$, we get that

$$||g^*||_\infty \leq ||\phi^*|| \, ||V||_1 \leq 2C\rho((||x_j^*||)_{j \geq 0}).$$

On the other hand, taking into account that $x \cdot e_j \in H_X^1$ for $j \geq 0$ and $x \in X$, we get that, for $0 \leq j \leq N$,

$$\widehat{g}^*(j)(x) = y_j^*(x) = \widehat{V}(j)\varphi^*(xe_j) = \widehat{V}(j)\varphi_0^*(xe_j) = x_j^*(x).$$

Thus, $\widehat{g}^*(j) = x_j^*$ for $0 \leq j \leq N$.

$(iii) \Rightarrow (i)$. Let $m \in FM$ and let f be an X-valued analytic trigonometric polynomial of degree N. For $j = 0, 1, \ldots, N$ pick $y_j^* \in X$ so that $||y_j^*|| = 1$ and $y_j^*(\widehat{f}(j)) = ||\widehat{f}(j)||$. Put $x_j^* = |m_j|y_j^*$ for $0 \leq j \leq N$ and $x_j^* = 0$ for $j > N$. Obviously, $\rho((||x_j^*||_{j \geq 0})) \leq \rho(m)$. We have that

$$\sum_{j=0}^{\infty} ||m_j \widehat{f}(j)|| = < f, g^* > \leq ||g^*||_\infty ||f||_1 \leq C\rho((||x_j^*||_{j \geq 0}))||f||_1 \leq$$

$$C\rho(m)||f||_1.$$

Hence m_X is $(1,1)$-bounded. The proof is complete. $\qquad\square$

Now we recall that a Banach space Y is *crudely finitely representable* in a Banach space X if there is $K \geq 1$ such that for every finite-dimensional subspace E of Y there is a linear operator $u : E \rightarrow X$ such that $||e|| \leq ||u(e)|| \leq K||e||$ for $e \in E$. As a useful example consider for $1 \leq p < \infty$ the space $H_X^p(D)$ of all X-valued analytic functions on the unit disk $D = \{z \in \mathbb{C}; |z| < 1\}$ such that for each $0 < r < 1$ the function $F_r \in H_X^p$ and $|||F|||_p = \sup\{||F_r||_p : 0 < r < 1\} < \infty$, where $F_r(t) = F(re^{it})$. Clearly H_X^p isometrically embeds into $H_X^p(D)$. Conversely, it is not hard to verify that $H_X^p(D)$ is crudely finitely representable in H_X^p.

Definition 4.13 clearly yields the following:

Corollary 4.21. *Every Banach space crudely finitely representable in a space of $(H^1 - \ell_1)$-Fourier type is an $(H^1 - \ell_1)$-Fourier type space.*

Let (μ, Ω) be a measure space and let X be a Banach space. By $L_X^1(\mu)$ we denote the space of X-valued Bochner μ integrable functions on Ω.

Proposition 4.22. *If X is an $(H^1 - \ell_1)$-Fourier type space, then so is $L_X^1(\mu)$.*

Proof. It is enough to show that $(\ell_1)_X$ is of $(H^1 - \ell_1)$-Fourier type, because for every measure space (μ, Ω), $L_X^1(\mu)$ is finitely representable in $(\ell_1)_X$.

Let $f = (f_k)$ be an $(\ell_1)_X$-valued analytic trigonometric polynomial. Then obviously each of the coordinates f_k is an X-valued analytic trigonometric polynomial. Hence, by the hypothesis on X there is $C > 0$, such that, for every $m \in FM$,

$$\sum_{j=0}^{\infty} ||m_j \widehat{f_k}(j)||_X \leq C\rho(m)||f_k||_X \quad (k = 0, 1, \dots).$$

Summing over all k, we get that

$$\sum_{j=0}^{\infty} ||m_j \widehat{f}(j)||_{(\ell_1)_X} = \sum_{j=0}^{\infty} \sum_{k=0}^{\infty} ||m_j \widehat{f_k}(j)||_X \leq \sum_{k=0}^{\infty} C\rho(m)||f_k||_X$$

$$= C\rho(m)||f||_{(\ell_1)_X}. \qquad \square$$

Another obvious but useful consequence of Definition 4.13 is the following:

Corollary 4.23. *Assume that a Banach space X satisfies the following condition:*

(∗)   *there is $C > 0$ such that for every $f \in H_X^1$ there is a complex valued function $\varphi \in H^1$ such that*

$$\|\widehat{f}(j)\| \leq |\widehat{\varphi}(j)| \text{ for } j = 0, 1, \ldots; \quad \|\varphi\|_1 \leq C\|f\|_1. \tag{4.20}$$

Then X is a $(H^1 - \ell_1)$-Fourier type space.

Theorem 4.24. *The space C_1 satisfies (∗) with $C = 1 + \epsilon$, for all $\epsilon > 0$, and, hence, it is a $(H^1 - \ell_1)$-Fourier type space.*

The proof of Theorem 4.24 is a consequence of the following result:

Theorem 4.25. (The noncommutative factorization theorem [45].) *Let $\epsilon > 0$. For every $f \in H_{C_1}^1$ there are g and h in $H_{C_2}^2$ such that*

$$f = g \cdot h, \quad (1 + \epsilon)\|f\|_{C_1,1} \geq \|g\|_{C_2,2}\|h\|_{C_2,2} \geq \|f\|_{C_1,1}.$$

Here $f = g \cdot h$ means that $f(t) = g(t) \cdot h(t)$ for $t \in \mathbb{R}$, i.e. at each point t the matrix $f(t)$ is the product of the matrix $h(t)$ with the matrix $g(t)$. Moreover,

$$\|f\|_{C_p,r} := \left((2\pi)^{-1} \int_{-\pi}^{\pi} \|f(t)\|_{C_p}^r dt \right)^{1/r} \quad \text{for } 1 \leq p < \infty \text{ and } 1 \leq r < \infty.$$

Remark: In fact Theorem 4.25 holds also for $\epsilon = 0$. This strong version is due to Sarason [85]. We need only the weaker ϵ-version of Sarason's theorem. In fact, we need only a much weaker version of Corollary 4.23 (see [84]), namely the following:

Corollary 4.26. *Assume that the analytic matrix Banach space X satisfies the following condition:*
(∗)   *there is $C > 0$ such that for every $A \in X$ there is a complex valued function $\varphi \in H^1$ such that*

$$\|A_j\| \leq |\widehat{\varphi}(j)| \text{ for } j = 0, 1, \ldots; \quad \|\varphi\|_1 \leq C\|A\|_X. \tag{4.21}$$

Then X has the following property: If there is a constant $K > 0$ such that, for a sequence of complex numbers (m_j) and $f \in H^1$ we have that $\left(\sum_{j=0}^{\infty} |m_j \widehat{f}j| \right) \leq K\|f\|_{H^1}$, then $\left(\sum_{j=0}^{\infty} \|m_j A_j\|_X \right) \leq K\|A\|_X$.

Next we present a proof of Theorem 4.24 by using Sarason's factorization theorem as in the Haagerup and Pisier's proof (see [45]):

Proof of Theorem 4.24. Let $f \in H^1_{C_1}$ and $g, h \in H^2_{C_2}$ satisfy the relation of Theorem 4.25. Put

$$\varphi(t) = \sum_{j=0}^{\infty} \sum_{k=0}^{j} ||\widehat{g}(k)||_{C_2} ||\widehat{h}(j-k)||_{C_2} e^{ijt}.$$

By using the inequality $||A \cdot B||_{C_1} \leq ||A||_{C_2} ||B||_{C_2}$ and that $f = g \cdot h$, we get that

$$|\widehat{\varphi}(j)| \geq \sum_{k=0}^{j} ||\widehat{g}(k) \cdot \widehat{h}(j-k)||_{C_1} \geq ||\sum_{k=0}^{j} \widehat{g}(k) \cdot \widehat{h}(j-k)||_{C_1} = ||\widehat{f}(j)||_{C_1}.$$

On the other hand, note that $\varphi = G \cdot H$, where

$$G = \sum_{j=0}^{\infty} ||\widehat{g}(k)||_{C_2} e^{ijt}, \quad H = \sum_{j=0}^{\infty} ||\widehat{h}(j)||_{C_2} e^{ijt}.$$

Next we observe that $||G||_2 = ||g||_{C_2,2}$ and $||H||_2 = ||h||_{C_2,2}$. Now using the Schwarz inequality and Theorem 4.25 we get that

$$||\varphi||_1 \leq ||G||_2 ||H||_2 = ||g||_{C_2,2} ||h||_{C_2,2} \leq (1+\epsilon) ||f||_{C_1,1}.$$

The proof is complete.

$\square$

Corollary 4.27. *The dual of $B(\ell_2)$ has $(H^1 - \ell_1)$-Fourier type.*

Proof.   We have $C_1^* = B(\ell_2)$. But it is well-known by the Local Reflexivity Principle [54] that the second dual of any Banach space is finitely representable in the space. Hence, the proof of the Corollary follows. $\square$

Now we give the proof of Theorem 4.25 as given in [45].

Proof of Theorem 4.25. First we recall the well-known fact (see [35]) that the projective tensor product $\ell_2 \widehat{\otimes} \ell_2$ may be isometrically identified with C_1. Here $\ell_2 \widehat{\otimes} \ell_2$ is the completion of algebraic tensor product $\ell_2 \otimes \ell_2$ under the norm

$$||u|| = \inf \{ \sum_i ||x_i|| \cdot ||y_i||; \ u = \sum_{i=1}^{\infty} x_i \otimes y_i \}.$$

Then, in the framework of tensor products the statement of Theorem 4.25 can be reformulated as follows:

Let $\epsilon > 0$. For any $f \in H^1_{\ell_2 \widehat{\otimes} \ell_2}$, there are sequences (g_k) and (h_k) in $H^2_{\ell_2}$ such that

$$\forall z \in D \quad f(z) = \sum_{k=1}^{\infty} g_k(z) \otimes h_k(z) \tag{4.22}$$

and

$$\|f\|_{H^1_{\ell_2 \hat\otimes \ell_2}} \le \sum_{k=1}^{\infty} \|g_k\|_{H^2_{\ell_2}} \|h_k\|_{H^2_{\ell_2}}. \tag{4.23}$$

Indeed, let (e_n) denote the canonical basis of ℓ_2. Let us denote by $(g_{ij}(z))$ and $(h_{ij}(z))$ the coefficients of the matrices $g(z)$ and $h(z)$ relative to the basis $(e_i \otimes e_j)$, and similarly for f.

We have, by the first relation in Theorem 4.25, that

$$f_{ij}(z) = \sum_k g_{ik}(z) h_{kj}(z)$$

and, hence,

$$f(z) = \sum f_{ij}(z) e_i \otimes e_j = \sum_k g_k(z) \otimes h_k(z),$$

where

$$g_k(z) = \sum_i g_{ik} e_i \text{ and } h_k(z) = \sum_j h_{kj} e_j.$$

This proves that (4.22) and (4.23) follow from the relations in Theorem 4.25. (The converse direction is also easy.)

Let us prove the relations (4.22) and (4.23).

We denote by P the linear subspace of $H^1_{C_1}$ formed by all the polynomials with coefficients in $\ell_2 \otimes \ell_2$.

Moreover, we denote by $\| \ \|_1$ the norm in $H^1_{C_1}$, and by $\| \ \|_2$ the norm in $H^2_{\ell_2}$.

Clearly, for every f in P there are polynomials with coefficients in ℓ_2 g_i, h_i such that

$$\forall z \in D \quad f(z) = \sum_{i=1}^n g_i(z) \otimes h_i(z).$$

We introduce a norm on P by

$$\|f\| = \inf \{ \sum_i^n \|g_i\|_2 \|h_i\|_2 \},$$

where the infimum runs over all possible representations.

Note that we obviously have that, $\|f\|_1 \le \|f\|$ and $\| \ \|$ is indeed a norm on P.

The main point of the proof of Theorem 4.25 is to check that actually this "new" norm $\|f\|$ coincides with $\|f\|_1$. Using duality, we will show that this follows rather directly from known results in the theory of vectorial Hankel operators due to S. Parott [70].

To explain this more precisely, we need to identify the dual spaces to P equipped with the norms $||\;||_1$ and $||\;||$.

Let us denote by Λ the space of all sequences $a = (a_n)_{n \geq 0}$ with $a_n \in B(\ell_2)$ such that the Hankel matrix $\mathcal{H}_a$ with entries $(\mathcal{H}_a)_{ij} = a_{i+j}$ ($i \geq 0$, $j \geq 0$) defines a bounded operator on $\ell_2(\ell_2) = \ell_2 \widehat{\otimes} \ell_2$. By definition, we set $||a|| = ||\mathcal{H}_a||$. (For a definition of a Hankel matrix and for some its properties see [65] and also Chapter 5.) Let us denote by X (resp. X_1) the normed space obtained by equipping P with the norm $||\;||$ (resp. $||\;||_1$).

We may introduce a duality between P and Λ as follows. Let (f_n) denote the Taylor coefficients of an element f in P. Then, for all a in Λ, we define

$$< a, f > := \sum_{n=0}^{\infty} < a_n, f_n > .$$

(Note that this sum is finite.)

With this duality, we have that

$$||a||_{X^*} = \sup < a, g \otimes h > = \sup \sum_{ij} (a_{ij} g_j, \overline{h}_i) = ||\mathcal{H}_a||,$$

where each of the above suprema runs over all g, h in P such that $||g||_2 \leq 1$ and $||h||_2 \leq 1$. (Of course we have that $||g||_2 = (\sum ||g_j||^2)^{1/2}$.)

This shows that Λ can be naturally identified isometrically with the dual of X.

Similarly, let us denote by $\widetilde{\Lambda}$ the space of all sequences $\alpha = (\alpha_n)_{n \in \mathbb{Z}}$ with $\alpha_n \in X^* \subset B(\ell_2)$ such that the matrix T_α defined by

$$(T_\alpha)_{ij} = \alpha_{i+j} \quad \forall i, j \in \mathbb{Z} \tag{4.24}$$

defines a bounded operator on $\ell_2(\mathbb{Z}, \ell_2)$. By definition, we set $||\alpha||_{\widetilde{\Lambda}} := ||T_\alpha||$.

Here again it is simple to check that $L_1(\mathbb{T}, C_1)^* = \widetilde{\Lambda}$ isometrically. Equivalently, this means that the natural mapping from $L_2(\ell_2) \widehat{\otimes} L_2(\ell_2)$ into $L_1(C_1)$ is a metric surjection. This can be viewed as a consequence of the identity $L_1(C_1) = L_1 \widehat{\otimes} C_1$ and the fact that every scalar function with L_1-norm 1 is the product of two functions with L_2-norm 1. Let us now return to our original problem to show that X coincides with X_1, or simply that $||f|| \leq ||f||_1$ for all f in P. To prove that it suffices to show that every a in the unit ball of X^* defines an element in the unit ball of X_1^{**}.

Equivalently, it is enough to show that for any $a = (a_n)_{n \geq 0}$ in the unit ball of $\Lambda = X^*$, there is an $\alpha = (\alpha_n)_{n \in \mathbb{Z}}$ in the unit ball of $L_1(C_1)^* = \tilde{\Lambda}$, which is such that $< a, f > = < \alpha, f >$ for all f in P. Clearly this means that $\alpha_n = a_n$ for all $n \geq 0$.

We have thus reduced our problem to the fact that every Hankel matrix with coefficients in $B(\ell_2)$ can be completed to a matrix with coefficients in $B(\ell_2)$ of the form (4.24) and of the same norm. This is precisely what Parrott shows in [70]. In fact, he gives an explicit inductive construction of the coefficients α_{-1}, α_{-2}, etc. which can be added to the sequence $a = (a_n)_{n \geq 0}$ in order to form an extended sequence with the desired property $||T_\alpha|| = ||\mathcal{H}_a||$.

This allows us to conclude that X and X_1 are identical. Since their completions must be also identical, we obtain the proof of the theorem. $\square$

Now we wish to prove Parrott's result, which we used previously.

In order to do this we state and prove some technical results concerning the completing matrix contractions. These results belong also to Parrott, but we follow the presentation from Peller's monograph [65].

Let $\mathcal{H}$, $\mathcal{K}$ be Hilbert spaces, A a bounded linear operator on $\mathcal{H}$, B a bounded linear operator from $\mathcal{K}$ to $\mathcal{H}$, and C a bounded linear operator from $\mathcal{H}$ to $\mathcal{K}$. The problem is to find out under which conditions there exists a bounded linear operator Z on $\mathcal{K}$ such that the operator

$$Q_Z = \begin{pmatrix} A \ B \\ C \ Z \end{pmatrix} \tag{4.25}$$

on $\mathcal{H} \oplus \mathcal{K}$ is a contraction, that is, $||Q_Z|| \leq 1$.

It is easy to see that if the problem is solvable, then the operators

$$\begin{pmatrix} A \\ C \end{pmatrix} \quad \text{and} \quad (A \ B) \tag{4.26}$$

from $\mathcal{H}$ to $\mathcal{H} \oplus \mathcal{K}$ and from $\mathcal{H} \oplus \mathcal{K}$ to $\mathcal{H}$, respectively, are contractions. It turns out that the converse is also true.

Theorem 4.28. *Let $\mathcal{H}$, $\mathcal{K}$ be Hilbert spaces and let, $A : \mathcal{H} \to \mathcal{H}$, $B : \mathcal{K} \to \mathcal{H}$, and $C : \mathcal{H} \to \mathcal{K}$ bounded linear operators. Then there is an operator $Z : \mathcal{K} \to \mathcal{K}$ for which the operator Q_Z defined by (4.25) is a contraction on $\mathcal{H} \oplus \mathcal{K}$ if and only if*

$$\left|\left| \begin{pmatrix} A \\ C \end{pmatrix} \right|\right| \leq 1 \quad \text{and} \quad ||(A \ B)|| \leq 1. \tag{4.27}$$

Next we describe all operators Z on $\mathcal{K}$ for which Q_Z is a contraction. In order to be able to state this description we need some preliminaries.

Lemma 4.29. *Let $\mathcal{H}$, $\mathcal{H}_1$, and $\mathcal{H}_2$ be Hilbert spaces, and let $T : \mathcal{H}_1 \to \mathcal{H}$ and $R : \mathcal{H}_2 \to \mathcal{H}$ be bounded linear operators. Then $TT^* \leq RR^*$ if and only if there exists a contraction $Q : \mathcal{H}_1 \to \mathcal{H}_2$ such that $T = RQ$.*

Proof.   Suppose that $T = RQ$ and $||Q|| \leq 1$. We have that

$$(TT^*x, x) = (RQQ^*R^*x, x) = (Q^*R^*x, Q^*R^*x)$$

$$= ||Q^*R^*x||^2 \leq ||R^*x||^2 = (RR^*x, x).$$

Conversely, assume that $TT^* \leq RR^*$. We define the operator L on Range R^* as follows:

$$L R^* x = T^* x, \quad x \in \mathcal{H}_2.$$

The inequality $TT^* \leq RR^*$ implies that L is well-defined on Range R^* and $||L|| \leq 1$ on Range R^*. We can extend L by continuity to the closure clos Range R^* and put

$$L|\mathrm{Ker}R = L|(\mathrm{Range}\,R^*)^{\perp} = 0.$$

Set $Q = L^*$. Clearly $T = RQ$. $\square$

For a contraction $A : \mathcal{H}_1 \to \mathcal{H}_2$ the *defect operator* D_A is defined on $\mathcal{H}_1$ by $D_A = (I - A^*A)^{1/2}$.

It is also convenient besides D_A to consider other operators $\mathcal{D}_A : \mathcal{H}_1 \to \widetilde{\mathcal{H}}$ such that $\mathcal{D}_A^* \mathcal{D}_A = I - A^*A$, where $\widetilde{\mathcal{H}}$ is a Hilbert space. In this case $\mathcal{D}_A = VD_A$ for some isometry V defined on clos Range D_A.

Lemma 4.30. *Let $\mathcal{H}$, $\mathcal{K}$, $\widetilde{\mathcal{H}}$ be Hilbert spaces, $A : \mathcal{H} \to \mathcal{H}$, $B : \mathcal{K} \to \mathcal{H}$ linear operators such that $||A|| \leq 1$. Let $\mathcal{D}_{A^*} : \mathcal{H} \to \widetilde{\mathcal{H}}$ be an operator such that $\mathcal{D}_A^* \mathcal{D}_{A^*} = I - AA^*$. Then*

$$|| \left(A \; B \right) || \leq 1 \tag{4.28}$$

if and only if $B = \mathcal{D}_{A^*}^* K$ for a contraction $K : \mathcal{K} \to \widetilde{\mathcal{H}}$.

Proof.   It is easy to see that (4.28) is equivalent to the fact that

$$\left(A \; B \right) \begin{pmatrix} A^* \\ B^* \end{pmatrix} \leq I_{\mathcal{H}},$$

which means that $AA^* + BB^* \leq I$ or, which is the same, $BB^* \leq \mathcal{D}_A^* \mathcal{D}_{A^*}$. By Lemma 4.29, this is equivalent to the fact that $B = \mathcal{D}_{A^*}^* K$ for some contraction $K : \mathcal{K} \to \widetilde{\mathcal{H}}$. The proof is complete. $\square$

Remark. The conclusion of Lemma 4.30 is valid if $\mathcal{D}_{A^*} = D_{A^*} = (I - AA^*)^{1/2}$. It is easy to see that we can choose a contraction $K : \mathcal{K} \to \mathcal{H}$ such that Range $K \subset$ clos Range$(I - AA^*)$ and $B = D_{A^*}K$. Clearly, such a contraction K is unique.

Lemma 4.31. *Let $\mathcal{H}$, $\mathcal{K}$, $\widetilde{\mathcal{H}}$ be Hilbert spaces and let $A : \mathcal{H} \to \mathcal{H}$, $C : \mathcal{H} \to \mathcal{K}$ be linear operators such that $\|A\| \leq 1$. Let $\mathcal{D}_A : \mathcal{H} \to \widetilde{\mathcal{H}}$ be an operator such that $\mathcal{D}_A^* \mathcal{D}_A = I - A^*A$. Then*

$$\left\| \begin{pmatrix} A \\ C \end{pmatrix} \right\| \leq 1 \tag{4.29}$$

if and only if $C = L\mathcal{D}_A$ for some contraction $L : \widetilde{\mathcal{H}} \to \mathcal{K}$.

Proof. The result follows from Lemma 4.30 since (4.29) is equivalent to the inequality

$$\| \begin{pmatrix} A^* & C^* \end{pmatrix} \| \leq 1. \qquad \square$$

Remark. As in Lemma 4.30 we can take $\mathcal{D}_A = D_A = (I - A^*A)^{1/2}$. Clearly, one can find a contraction $L : \mathcal{H} \to \mathcal{K}$ such that

$$L|(\text{Range}(I - A^*A))^{\perp} = 0$$

and $C = LD_A$. As in Lemma 4.30 it is easy to see that such a contraction L is unique.

Now we are in a position to state the description of those operators $Z : \mathcal{K} \to \mathcal{K}$ for which the operator Q_Z defined by (4.25) is a contraction. As we have already observed, the operators in (4.26) are contractions. Therefore (see the Remarks after Lemmas 4.30 and 4.31) there exist unique contractions $K : \mathcal{K} \to \mathcal{H}$ and $L : \mathcal{H} \to \mathcal{K}$ such that

$$\text{Range} K \subset \text{clos Range}(I - AA^*), \quad B = D_{A^*}K, \tag{4.30}$$

$$L|(\text{Range}(I - A^*A))^{\perp} = 0, \quad C = LD_A. \tag{4.31}$$

Theorem 4.32. *Let $\mathcal{H}$, $\mathcal{K}$ be Hilbert spaces, $A : \mathcal{H} \to \mathcal{H}$, $B : \mathcal{K} \to \mathcal{H}$, and $C : \mathcal{H} \to \mathcal{K}$ bounded linear operators satisfying (4.27). Let $K : \mathcal{K} \to \mathcal{H}$ and $L : \mathcal{H} \to \mathcal{K}$ be the operators satisfying (4.30) and (4.31). If $Z : \mathcal{K} \to \mathcal{K}$ is a bounded linear operator, then the operator Q_Z, defined by (4.25), is a contraction if and only if Z admits a representation*

$$Z = -LA^*K + D_{L^*}MD_K, \tag{4.32}$$

where M is a contraction on $\mathcal{K}$.

Note that we may always assume that

$$M|(\text{Range} D_K)^{\perp} = 0 \quad \text{and} \quad \text{Range} M \subset \text{clos Range} D_{L^*}. \qquad (4.33)$$

If these two conditions are satisfied, then Z determines M uniquely, and so the contractions M satisfying (4.33) parametrize the solutions Z.

It is easy to see that Theorem 4.28 follows from Theorem 4.32. Indeed, we can always take $M = 0$. To prove Theorem 4.32, we need one more lemma.

Lemma 4.33. *Let A, B be as above and let $K : \mathcal{K} \to \mathcal{H}$ be an operator satisfying (4.30). Then the operator*

$$\mathcal{D}_{(AB)} = \begin{pmatrix} D_A & -A^*K \\ 0 & D_K \end{pmatrix} \qquad (4.34)$$

satisfies

$$\mathcal{D}^*_{(AB)} \mathcal{D}_{(AB)} = I_{\mathcal{H} \oplus \mathcal{K}} - \begin{pmatrix} A^* \\ B^* \end{pmatrix} \begin{pmatrix} A & B \end{pmatrix}.$$

Proof.   We have that

$$\begin{pmatrix} D_A & -A^*K \\ 0 & D_K \end{pmatrix}^* \begin{pmatrix} D_A & -A^*K \\ 0 & D_K \end{pmatrix} + \begin{pmatrix} A^* \\ B^* \end{pmatrix} \begin{pmatrix} A & B \end{pmatrix}$$

$$= \begin{pmatrix} D_A & 0 \\ -K^*A & D_K \end{pmatrix} \begin{pmatrix} D_A & -A^*K \\ 0 & D_K \end{pmatrix} + \begin{pmatrix} A^* \\ B^* \end{pmatrix} \begin{pmatrix} A & B \end{pmatrix}$$

$$= \begin{pmatrix} D_A^2 & -D_A A^*K \\ -K^*A D_A & K^*AA^*K + D_K^2 \end{pmatrix} + \begin{pmatrix} A^*A & A^*B \\ B^*A & B^*B \end{pmatrix}$$

$$= \begin{pmatrix} I_{\mathcal{H}} - A^*A & -D_A A^*K \\ -K^*A D_A & K^*K + D_K^2 - K^*D_{A^*} D_{A^*} K \end{pmatrix}$$

$$\quad + \begin{pmatrix} A^*A & A^*B \\ B^*A & B^*B \end{pmatrix}.$$

Let us show that $\bar{D}_A A^* = A^* D_{A^*}$. We have that $(I - A^*A)A^* = A^*(I - AA^*)$. It follows that $\phi(I - A^*A)A^* = A^*\phi(I - AA^*)$ for any polynomial ϕ, so the same equality holds for any continuous function ϕ. If we take $\phi(t) = t^{1/2}$, $t \geq 0$, we obtain that $D_A A^* = A^* D_{A^*}$. Similarly, $D_{A^*} A = AD_A$. Consequently,

$$\mathcal{D}^*_{(AB)} \mathcal{D}_{(AB)} + \begin{pmatrix} A^* \\ B^* \end{pmatrix} \begin{pmatrix} A & B \end{pmatrix}$$

$$= \begin{pmatrix} I_{\mathcal{H}} - A^*A & -A^*B \\ -B^*A & I - B^*B \end{pmatrix} + \begin{pmatrix} A^*A & A^*B \\ B^*A & B^*B \end{pmatrix} = \begin{pmatrix} I_{\mathcal{H}} & 0 \\ 0 & I_{\mathcal{H}} \end{pmatrix}. \qquad \square$$

Proof of Theorem 4.32. Suppose that $\|Q_Z\| \leq 1$. By Lemma 4.29

$$\begin{pmatrix} C & Z \end{pmatrix} = \begin{pmatrix} X & Y \end{pmatrix} \mathcal{D}_{(AB)},$$

where $\begin{pmatrix} X & Y \end{pmatrix}$ is a contraction from $\mathcal{H} \oplus \mathcal{K}$ to $\mathcal{K}$ and $\mathcal{D}_{(AB)}$ is defined by (4.34). Then $C = XD_A$. Let P be the orthogonal projection from $\mathcal{H}$ onto clos Range$(I - A^*A)$. Put $\widetilde{X} = XP$. Clearly, $C = \widetilde{X}D_A$ and, by the Remark after Lemma 4.31, we have that $\widetilde{X} = L$. It is easy to see that

$$\begin{pmatrix} C & Z \end{pmatrix} = \begin{pmatrix} L & Y \end{pmatrix} \mathcal{D}_{(AB)}. \tag{4.35}$$

Clearly,

$$\begin{pmatrix} L & Y \end{pmatrix} = \begin{pmatrix} X & Y \end{pmatrix} \begin{pmatrix} P & 0 \\ 0 & I_{\mathcal{H}} \end{pmatrix},$$

which proves that $\begin{pmatrix} L & Y \end{pmatrix}$ is a contraction. Then, by Lemma 4.30, the operator Y admits a representation $Y = D_{L^*}M$ for a contraction M on $\mathcal{K}$. Formula (4.32) follows now immediately from (4.35).

Suppose that Z satisfies (4.32), where M is a contraction on $\mathcal{K}$. Then it is easy to see that

$$\begin{pmatrix} A & B \\ C & Z \end{pmatrix} = \begin{pmatrix} I & 0 & 0 \\ 0 & L & D_{L^*} \end{pmatrix} \begin{pmatrix} A & D_{L^*} & 0 \\ D_A & -A^* & 0 \\ 0 & 0 & M \end{pmatrix} \begin{pmatrix} I & 0 \\ 0 & K \\ 0 & D_K \end{pmatrix}. \tag{4.36}$$

The result follows from the fact that all factors on the right-hand side of (4.36) are contractions, which is a consequence of the following lemma:

Lemma 4.34. *Let T be a contraction on a Hilbert space. Then the operator*

$$\begin{pmatrix} T & D_{T^*} \\ D_T & -T^* \end{pmatrix}$$

is unitary.

Proof. We have

$$\begin{pmatrix} T & D_{T^*} \\ D_T & -T^* \end{pmatrix}^* \begin{pmatrix} T & D_{T^*} \\ D_T & -T^* \end{pmatrix} = \begin{pmatrix} T^*T + D_T^2 & T^*D_{T^*} - D_TT^* \\ D_{T^*}T - TD_T & D_{T^*}^2 + TT^* \end{pmatrix}.$$

It has been shown in the proof of Lemma 4.33 that $T^*D_{T^*} = D_TT^*$ and $TD_T = D_{T^*}T$. Thus,

$$\begin{pmatrix} T & D_{T^*} \\ D_T & -T^* \end{pmatrix}^* \begin{pmatrix} T & D_{T^*} \\ D_T & -T^* \end{pmatrix} = \begin{pmatrix} I & 0 \\ 0 & I \end{pmatrix}.$$

Similarly,

$$\begin{pmatrix} T & D_{T^*} \\ D_T & -T^* \end{pmatrix} \begin{pmatrix} T & D_{T^*} \\ D_T & -T^* \end{pmatrix}^* = \begin{pmatrix} I & 0 \\ 0 & I \end{pmatrix}.$$

The proof is complete. □

Remark. The same results are valid if A is an operator from $\mathcal{H}_1$ to $\mathcal{H}_2$, B is an operator from $\mathcal{K}_1$ to $\mathcal{K}_2$, C is an operator from $\mathcal{H}_1$ to $\mathcal{K}_2$, and Z is an operator from $\mathcal{K}_1$ to $\mathcal{K}_2$, where $\mathcal{H}_1$, $\mathcal{H}_2$, $\mathcal{K}_1$, $\mathcal{K}_2$ are Hilbert spaces. The proof given above works also in this more general situation.

Remark. It is clear that if we replace in Theorem 4.28 $\mathcal{H}$ by $\mathcal{K}$ and conversely we get that the matrix with operators as its entries

$$\begin{pmatrix} Z & C \\ B & A \end{pmatrix}$$

is a contraction on $\mathcal{K} \oplus \mathcal{H}$ if and only if $\begin{pmatrix} B & A \end{pmatrix}$ and $\begin{pmatrix} C \\ A \end{pmatrix}$ are contractions.

Now let us present the Parrott's argument to complete the proof of Theorem 4.25 (see [70]): Consider the *block Hankel matrix* (that is a matrix having operators as its entries) $\Gamma_\Omega = \{\Omega_{j+k}\}_{j,k \geq 0}$, where $\Omega = \{\Omega_j\}_{j \geq 0}$ is a sequence of bounded linear operators from $\mathcal{H}$ to $\mathcal{K}$, for two separable Hilbert spaces $\mathcal{H}$, $\mathcal{K}$, and the matrix Γ_Ω is of the form:

$$\begin{pmatrix} \Omega_0 & \Omega_1 & \Omega_2 & \Omega_3 & \cdots \\ \Omega_1 & \Omega_2 & \Omega_3 & \Omega_4 & \cdots \\ \Omega_2 & \Omega_3 & \Omega_4 & \vdots & \cdots \\ \Omega_3 & \Omega_4 & \vdots & & \\ \vdots & \vdots & & & \end{pmatrix}.$$

Now we wish to construct a block Hankel matrix $\widetilde{\Gamma_\Omega}$ of the form

$$\begin{pmatrix} \vdots & \vdots & \vdots & \cdots \\ \Omega_{-2} & \Omega_{-1} & \Omega_0 & \cdots \\ \Omega_{-1} & \Omega_0 & \Omega_1 & \cdots \\ \Omega_0 & \Omega_1 & \Omega_2 & \cdots \\ \Omega_1 & \Omega_2 & \Omega_3 & \vdots \\ \vdots & \vdots & \vdots & \end{pmatrix}$$

such that $\|\widetilde{\Gamma_\Omega}\| = \|\Gamma_\Omega\|$.

We construct $\widetilde{\Gamma_\Omega}$ inductively, first constructing a Hankel matrix

$$\Gamma_{-1} = \begin{pmatrix} Z & \Omega_0 & \Omega_1 & \Omega_2 & \cdots \\ \Omega_0 & \Omega_1 & \Omega_2 & \Omega_3 & \cdots \\ \Omega_1 & \Omega_2 & \Omega_3 & \Omega_4 & \cdots \\ \Omega_2 & \Omega_3 & \Omega_4 & \Omega_5 & \vdots \\ \vdots & \vdots & \vdots & \vdots & \ddots \end{pmatrix}$$

such that $||\Gamma_{-1}|| = ||\Gamma_\Omega||$. This of course involves the choice of $\Omega_{-1} = Z$. Given that this construction is always possible, successive iterations will produce $\Omega_{-2}, \Omega_{-3}, \ldots$, and $\widetilde{\Gamma_\Omega}$. To choose Ω_{-1} we write

$$A = \begin{pmatrix} \Omega_1 & \Omega_2 & \Omega_3 & \ldots \\ \Omega_2 & \Omega_3 & \Omega_4 & \ldots \\ \Omega_3 & \Omega_4 & \Omega_5 & \ldots \\ \vdots & \vdots & \vdots & \ddots \end{pmatrix}, \quad B = \begin{pmatrix} \Omega_0 \\ \Omega_1 \\ \Omega_2 \\ \vdots \end{pmatrix},$$

$$C = \begin{pmatrix} \Omega_0 & \Omega_1 & \Omega_2 & \ldots \end{pmatrix}.$$

The matrix Γ_{-1} can be identified with the matrix

$$\begin{pmatrix} Z & C \\ B & A \end{pmatrix}.$$

Clearly,

$$|| (B\ A) || = ||\Gamma_\Omega||, \quad \left\|\begin{pmatrix} C \\ A \end{pmatrix}\right\| = ||\Gamma_\Omega||.$$

By Theorem 4.28 and the last two remarks, there exists an operator Z such that

$$\left\|\begin{pmatrix} Z & C \\ B & A \end{pmatrix}\right\| = ||\Gamma_\Omega||.$$

Now put $\Omega_{-1} = Z$. We have

$$||\Gamma_{-1}|| = ||\Gamma_\Omega||.$$

The proof of Theorem 4.25 is complete. $\qquad\qquad\square$

4.5 A characterization of the space T_1

In the book of M. Pavlović ([68] page 96) there is the following beautiful characterization of functions belonging to the Hardy space H^1:

Pavlović Theorem *For a function f, which is analytic in D, the following assertions are equivalent:*

a) $f \in H^1$;

b) $\displaystyle\sup_n \frac{1}{a_n} \sum_{j=0}^{n} \frac{1}{j+1} ||s_j f|| < \infty$;

$$c) \ \sup_n ||P_n f|| < \infty.$$

Here, for a function f analytic in D let

$$P_n f = \frac{1}{a_n} \sum_{j=0}^{n} \frac{1}{j+1} s_j f, \ \text{where} \ a_n = \sum_{j=0}^{n} \frac{1}{j+1} \quad (n = 0, 1, 2, \dots)$$

and $s_j f$ are the partial sums of the Taylor series of f.

An analogue of this result using the vector-valued Hardy inequality given in Section 4.4 is also true and is presented below.

More precisely, we have the following result:

Theorem 4.35. *Let $A \in B(\ell_2)$ be an upper triangular matrix. The following assertions are equivalent:*

$$a) \ A \in T_1;$$

$$b) \ \sup_n \frac{1}{a_n} \sum_{j=0}^{n} \frac{1}{j+1} ||s_j A|| < \infty;$$

$$c) \ \sup_n ||P_n A|| < \infty.$$

Here

$$P_n A = \frac{1}{a_n} \sum_{j=0}^{n} \frac{1}{j+1} s_j A, \ \text{where} \ a_n = \sum_{j=0}^{n} \frac{1}{j+1} \quad (n = 0, 1, 2, \dots)$$

and $s_j A = \sum_{k=0}^{j} A_k$.

Proof. Obviously $b) \Rightarrow c)$.

$a) \Rightarrow b)$. Let $A \in T_1$, and for fixed $n \geq 2$, $w \in D$, and $r = 1 - \frac{1}{n} < 1$, define the matrix-valued function $g(z) = (1 - rz)^{-1} [A * C(rwz)] \ (|z| \leq 1)$, where $C(z)$ is the Toeplitz matrix corresponding to the function $\frac{1}{1-z}$ for each $z \in D$.

Then we have that

$$g(z) = \left(\sum_{k=0}^{\infty} A_k r^k w^k z^k \right) \left(\sum_{l=0}^{\infty} r^l z^l \right) = \sum_{k,l=0}^{\infty} A_k w^k r^{k+l} z^{k+l} =$$

$$\sum_{m=0}^{\infty}\left(\sum_{k=0}^{m}A_k w^k\right)r^m z^m = \sum_{m=0}^{\infty}s_m(A*C(w))r^m z^m.$$

Hence $\widehat{g}(m) = s_m(A*C(w))r^m$, $m = 0,1,2,\dots$

It is well-known (and easy to see) that

$$\|s_m A\|_{T_1} \le C\ln(m+1)\|A\|_{T_1} \quad \forall A \in T_1 \text{ and } m \in \mathbb{N}, \tag{4.37}$$

where $C > 0$ is an absolute constant.

Moreover, $g \in H_{C_1}^1$ since, by (4.37), we have that

$$\|s_m(A*C(w))\|_{T_1} \le \frac{1}{1-|w|}\|s_m A\|_{T_1} \le \frac{C\ln(m+1)}{1-|w|} \quad \forall m \in \mathbb{N} \text{ and } |w| < 1.$$

Therefore

$$\sum_{m=0}^{\infty}\|s_m(A*C(w))\|_{T_1}r^m \le \frac{C\sum_{m=0}^{\infty}r^m\ln(m+1)}{1-|w|} < \infty.$$

We conclude that

$$\sum_{j=0}^{\infty}\frac{1}{j+1}\|s_j(A*C(w))\|_{T_1}r^j = \sum_{j=0}^{\infty}\frac{1}{j+1}\|\widehat{g}(j)\|_{T_1}$$

$$\le \text{ (by Corollary 4.14 for } X = T_1)$$

$$\le C\|g\|_{H_{T_1}^1} = \frac{\|A*C(rwe^{it})\|_{T_1}}{|1-re^{it}|} \quad \text{for all } t \in [0,2\pi).$$

Since $r^j = (1-\frac{1}{n})^j \ge c \; \forall 0 \le j \le n$, where $c > 0$ is an absolute constant, we have:

$$\sum_{j=0}^{n}\frac{1}{j+1}\|s_j(A*C(w))\|_{T_1} \le C\int_0^{2\pi}\|g(re^{it})\|_{T_1}\frac{dt}{2\pi}$$

$$= C\int_0^{2\pi}\frac{\|A*C(rwe^{it})\|_{T_1}}{|1-re^{it}|}\frac{dt}{2\pi}.$$

Integrating this inequality over the circle $|w| = 1$ and since $s_j(A*C(w)) = s_j(A)*C(w)$, we find, using $\lim_{w\to e^{i\theta}}\|s_j(A)*C(w)\|_{T_1} = \|s_j A*C(e^{i\theta})\|_{T_1} \; \forall j$, that

$$\sum_{j=0}^{n}\frac{1}{j+1}\int_0^{2\pi}\|s_j A*C(e^{i\theta})\|_{T_1}\frac{d\theta}{2\pi} \le \frac{\pi}{c}\int_0^{2\pi}\int_0^{2\pi}\frac{\|A*C(re^{i(\theta+t)})\|_{T_1}}{|1-re^{it}|}\frac{dt}{2\pi}\frac{d\theta}{2\pi}$$

$$= \frac{\pi}{c} \int_0^{2\pi} \left(\int_0^{2\pi} ||A * P_r(t+\theta)||_{T_1} \frac{d\theta}{2\pi} \right) \frac{dt}{2\pi|1 - re^{it}|}$$

$$\leq \frac{\pi}{c} ||A||_{T_1} \frac{1}{2\pi} \int_0^{2\pi} \frac{dt}{|1 - re^{it}|} \leq C||A||_{T_1} \ln n,$$

where $P_r(t+\theta)$ is the usual Poisson kernel on the unit circle and $C > 0$ is an absolute constant.

However, denoting by E_θ the Toeplitz matrix corresponding to δ_θ (the Dirac measure concentrated in θ) we have that

$$\frac{1}{2\pi} \int_0^{2\pi} ||s_j A * C(e^{i\theta})||_{T_1} d\theta = \frac{1}{2\pi} \int_0^{2\pi} ||s_j A * E_\theta||_{T_1} d\theta.$$

Since

$$||B||_{T_1} = ||B * E_\theta * E_{2\pi-\theta}||_{T_1} \leq ||B * E_\theta||_{T_1} ||E_{2\pi-\theta}||_{M(\ell_2)} = ||B * E_\theta||_{T_1}$$

$$\leq ||B||_{T_1} ||E_\theta||_{M(\ell_2)} = ||B||_{T_1} \ \forall \theta \in [0, 2\pi]$$

it follows that

$$\sum_{j=0}^n \frac{1}{j+1} ||s_j A||_{T_1} \leq \sum_{j=1}^n \frac{1}{j+1} \int_0^{2\pi} ||s_j A * C(e^{i\theta})||_{T_1} \frac{d\theta}{2\pi} \leq C||A||_{T_1} \ln n,$$

that is

$$\frac{1}{a_n} \sum_{j=0}^n \frac{1}{j+1} ||s_j A||_{T_1} \leq C_1 ||A||_{T_1}$$

and b) holds.

$c) \Rightarrow a)$ It is clear that if A is a finite matrix, then

$$||A||_{C_1} \leq \sup_n ||P_n A||_{C_1}.$$

Now assume that A is any matrix such that $\sup_n ||P_n A||_{C_1} < \infty$. Let E_m be the canonical projection which projects a matrix to its submatrix of order m at the left upper corner. Since P_n and E_m commute, we find that

$$\sup_m \sup_n ||P_n E_m A||_{C_1} < \infty.$$

By the preceding remark, we have that

$$\sup_m ||E_m A||_{C_1} \leq \sup_m \sup_n ||P_n E_m A||_{C_1} < \infty;$$

whence $A \in C_1$ and

$$||A||_{C_1} \leq \sup_n ||P_n A||_{C_1}.$$

This inequality holds without the assumption that A is upper triangular. The proof is complete. □

A simple consequence of Theorem 4.35 is the following:

Corollary 4.36. *If $A \in T_1$, then*

$$\lim_n \frac{1}{a_n} \sum_{j=0}^{n} \frac{1}{j+1} ||A - s_j A|| = 0 \tag{4.38}$$

and, consequently,

$$\lim_n \frac{1}{a_n} \sum_{j=0}^{n} \frac{1}{j+1} ||s_j A|| = ||A||. \tag{4.39}$$

Proof. Obviously (4.38) holds if A is a finite matrix. Since finite matrices are dense in T_1 the proof of (4.38) follows. The second assertion follows immediately from (4.38). $\qquad\square$

We remark that B. Smith [89] proved in 1983 the relation (4.39) for $f \in H^1$ instead of $A \in T_1$, and this in fact, motivated Pavlović to give his theorem.

As a consequence of this theorem we have that

Corollary 4.37. *If $A \in T_1$, then $\liminf_{n \to \infty} ||A - s_n A||_{T_1} = 0$.*

Of course the last three results are matrix versions of some theorems concerning the strong convergence in H^1. (See [68].)

4.6 An extension of Shields's inequality

In this section we present an extension of the matrix version of Shields's inequality.

The results concerning the functions and analytic measures on $\mathbb{T}$ were obtained by C. McGehee, L. Pigno and B. Smith in 1981 (see [63]).

In particular, the above mentioned authors proved the Littlewood conjecture [40] from 1948.

In what follows we denote by $M(\mathbb{T})$ the usual convolution algebra of Borel measures on $\mathbb{T}$.

First we present the following generalization of the classical Hardy inequality:

Theorem 4.38. *There is a real number $C > 0$ such that, given any set $S = \{n_1 < n_2 < \ldots\} \subset \mathbb{Z}$ and $\mu \in M(\mathbb{T})$, with supp $\hat{\mu} \subset S$, the following* *inequality holds:*

$$\sum_{k=1}^{\infty} \frac{|\hat{\mu}(n_k)|}{k} \leq C||\mu||.$$

A direct consequence of Theorem 4.38 is the following result:

Corollary 4.39. *If $p(\theta) = \sum_{k=1}^{N} c_k e^{in_k\theta}$ where $\{n_1 < n_2 < \cdots < n_k\} \subset \mathbb{Z}$ and $|c_k| \geq 1$ for all k, then*

$$||p||_1 \geq \frac{1}{C} \log N.$$

The following lemma will be quite useful in the sequel.

Lemma 4.40. *Let $h \in L^2(\mathbb{T})$ and suppose that $0 \leq \Re h \in L^\infty(\mathbb{T})$ and supp $\hat{h} \subset \mathbb{Z}$. Then*

(a) $e^{-h} \in L^\infty(\mathbb{T})$, supp $\widehat{\{e^{-h}\}} \subset \mathbb{Z}^- = \{n \in \mathbb{Z}; \, n < 0\}$
    and
(b) $||e^{-h} - 1||_2 \leq ||h||_2$.

Proof.   To prove (b) we notice that, for all $z \in \mathbb{C}$ with $\Re z \geq 0$,

$$\left| \frac{e^{-z} - 1}{z} \right| \leq 1.$$

Therefore, since $\Re h \geq 0$, it follows that

$$||e^{-h} - 1||_2 \leq ||h||_2. \qquad \square$$

Proof of Theorem 4.38. Let $S = \{n_1 < n_2 \ldots\} \subset \mathbb{Z}$ and $\mu \in M(\mathbb{T})$ be given with supp $\hat{\mu} \subset S$. Put $S_0 = \{n_1\}$, $S_1 = \{n_2 < n_3 < n_4 < n_5\}$, $S_2 = \{n_6 < \cdots < n_{21}\}$; continuing in this fashion we obtain disjoint sets S_j satisfying:

$$S = \bigcup_{j=0}^{\infty} S_j$$

and

$$\text{card } S_j = 4^j \quad (j = 0, 1, 2, \ldots). \tag{4.40}$$

Define now trigonometric polynomials f_j $(j = 0, 1, 2, \ldots)$ by requiring that

$$\hat{f}_j = 0 \quad \text{off } S_j; \tag{4.41}$$

$$|\widehat{f_j}(n)| = 4^{-j} \quad \text{if } n \in S_j; \tag{4.42}$$

$$\widehat{f_j}\widehat{\mu} \geq 0. \tag{4.43}$$

Put $|f_j| = \sum_{-\infty}^{\infty} c_n e^{in\theta}$ and define

$$h_j(\theta) := \frac{1}{4}\{c_0 + 2\sum_{-\infty}^{-1} c_n e^{in\theta}\}.$$

Moreover, observe that, via the conditions (4.40), (4.41), (4.42), we have that

$$\|f_j\|_2 = 2^{-j}, \tag{4.44}$$

so that

$$\|h_j\|_2 \leq \frac{\sqrt{2}}{4}\|f_j\|_2 = \frac{\sqrt{2}}{4}2^{-j} < 3 \cdot 2^{-j-3}. \tag{4.45}$$

Let $F_0 = (1/5)f_0$ and define inductively,

$$F_{j+1} = F_j e^{-h_{j+1}} + \frac{1}{5}f_{j+1} \quad (j = 0, 1, 2, \dots).$$

Notice that since $\Re h_j = (1/4)|f_j|$, we have that $e^{-h_j} \in L^\infty(\mathbb{T})$ because $|f_j| \leq 1$ from conditions (4.40) through (4.42). Thus $F_j \in L^\infty(\mathbb{T})$ and we find, according to (4.40) and the inequality $e^{-x/4} + x/5 \leq 1$ for $0 \leq x \leq 1$ that

$$\|F_j\|_\infty \leq 1 \quad (j = 0, 1, 2, \dots).$$

Moreover, since supp $\widehat{h_j} \subset \mathbb{Z}^-$, we have that

$$\text{supp}\{e^{h_j}\}\widehat{} \subset \mathbb{Z}^- \tag{4.46}$$

via part (a) of Lemma 4.40. We now claim that, for any m,

$$|\widehat{F_m}(n) - \frac{1}{5}\widehat{f_j}(n)| \leq \frac{1}{10}|\widehat{f_j}(n)| \quad \text{if } n \in S_j \text{ and } j \leq m. \tag{4.47}$$

In order to prove the above assertion we first observe that

$$F_m = \frac{1}{5}f_0 e^{-\sum_1^m h_k} + \frac{1}{5}f_1 e^{-\sum_2^m h_k} + \dots$$

$$+\frac{1}{5}f_{m-1}e^{-h_m} + \frac{1}{5}f_m;$$

as a consequence of (4.41) and (4.46). Hence, if $n \in S_j$ and $j < m$, then

$$\widehat{F_m}(n) - \frac{1}{5}\widehat{f_j}(n) = \left\{\frac{f_j}{5}\left[e^{-\sum_{j+1}^m h_k} - 1\right]\right\}\widehat{}(n)$$

$$+ \left\{ \frac{f_{j+1}}{5} \left[e^{-\sum_{j+2}^{m} h_k} - 1 \right] \right\}^{\widehat{}} (n)$$

$$+ \cdots + \left\{ \frac{f_{m-1}}{5} \left[e^{-h_m} - 1 \right] \right\}^{\widehat{}} (n).$$

It is now obvious from part (b) of Lemma 4.40 and the Cauchy-Schwarz inequality that

$$|\widehat{F}_m(n) - \frac{1}{5}\widehat{f}_j(n)| \le \{||f_j||_2|| \sum_{j+1}^{m} h_k||_2 + ||f_{j+1}||_2|| \sum_{j+2}^{m} h_k||_2 + \cdots$$

$$+ ||f_{m-1}||_2||h_m||_2\}.$$

Thus, in view of the inequalities (4.44) and (4.45), we obtain that

$$|\widehat{F}_m(n) - \frac{1}{5}\widehat{f}_j(n)| \le \frac{3}{10}(4^{-j-1} + 4^{-j-2} + \dots) = \frac{1}{10}4^{-j} = \frac{1}{10}|\widehat{f}_j(n)|.$$

Let $j > 0$ and suppose that $n_k \in S_j$. Then $3k > 4^j$ and, hence,

$$|\widehat{f}_j(n_k)| = 4^{-j} > \frac{1}{3k}. \tag{4.48}$$

Notice that

$$\Re(\widehat{F}_m\widehat{\mu})(n_k) \ge \frac{1}{10}\widehat{f}_j(n_k)\widehat{\mu}(n_k) \tag{4.49}$$

because of the inequalities (4.43) and (4.47). Moreover, as a consequence of the inequalities (4.43), (4.48) and (4.49), we see that if $n_k \in S_j$, $j \le m$, then

$$\Re(\widehat{F}_m\widehat{\mu})(n_k) \ge \frac{1}{30}|\widehat{\mu}(n_k)|. \tag{4.50}$$

Put $B_m = S_0 \cup S_1 \cup \cdots \cup S_m$ and assume for the moment that μ is a trigonometric polynomial on $\mathbb{T}$. Then, on the one hand

$$|(F * \mu)(0)| \le ||\mu||_1$$

because $||F_m||_\infty \le 1$, while on the other hand

$$|(F_m * \mu)(0)| = | \sum_{n \in B_m} \widehat{F}_m(n)\widehat{\mu}(n)|.$$

Hence, the inequality (4.50) permits us to conclude that, for all trigonometric polynomials μ with supp $\widehat{\mu} \subset S$,

$$\sum_1^\infty \frac{|\widehat{\mu}(n_k)|}{k} \le 30||\mu||_1. \tag{4.51}$$

A standard approximate identity argument implies the inequality (4.51) for all $\mu \in M(\mathbb{T})$ with supp $\widehat{\mu} \subset S$. The proof is complete. $\square$

Proof of Corollary 4.39. Given $p(\theta) = \sum_{k=1}^{N} c_k e^{in_k\theta}$ $(n_1 < n_2 < \cdots < n_N)$, it follows from Theorem 4.38 that

$$\sum_{k=1}^{N} \frac{|\widehat{p}(n_k)|}{k} = \sum_{k=1}^{N} \frac{|c_k|}{k} \leq C\|p\|_1.$$

Since $|c_k| \geq 1$ for all k we see that $\log N \leq \sum_{k=1}^{N} \frac{1}{k} \leq C\|p\|_1$. $\square$

Obviously, we have also proved the following version of Corollary 4.39:

Let $\mu \in M(\mathbb{T})$ and suppose for $\{n_1 < n_2 < \cdots < n_N\} \subset \mathbb{Z}^-$ we have that $\widehat{\mu}(n_k) = c_k$, where $|c_k| \geq 1$ for $k = 1, 2, \ldots, N$ and $\widehat{\mu}(n) = 0$ for all other $n \in \mathbb{Z}^-$. Then

$$\|\mu\| \geq \frac{1}{C} \log N.$$

Corollary 4.41. *Given $S = \{n_1 < n_2 < \ldots\} \subset \mathbb{Z}$, and $\{c_k\}_1^{\infty} \subset \mathbb{C}$ such that $|c_k| \leq 1/k$, there is an $F \in L^{\infty}(\mathbb{T})$, $\|F\|_{\infty} \leq C$, such that $\widehat{F}(n_k) = c_k$ for all k.*

Proof. Given $S = \{n_1 < n_2 < \ldots\} \subset \mathbb{Z}$, put $L_S^1(\mathbb{T}) = \{g \in L^1(\mathbb{T}) :$ supp $\widehat{g} \subset S\}$; define

$$\Lambda(g) := \sum_{1}^{\infty} c_k \widehat{g}(n_k) \qquad (g \in L_S^1(\mathbb{T})),$$

where $\{c_k\}_1^{\infty} \subset \mathbb{C}$ is fixed and $|c_k| \leq 1/k$ for all k. It follows from Theorem 4.38 that Λ is a bounded linear functional on $L_S^1(\mathbb{T})$ such that $\|\Lambda\| \leq C$. Extend Λ to $L^1(\mathbb{T})$ by the Hahn-Banach theorem. We have thus obtained an $F \in L^{\infty}(\mathbb{T})$, $\|F\|_{\infty} \leq C$, such that

$$\widehat{F}(n_k) = c_k \quad \text{for all } k.$$

$\square$

We recall that Theorem 4.24 implies that C_1 is of $(H^1 - \ell_1)$-Fourier type. On the other hand, by Theorem 4.38, it follows that the sequence $\{a_n\}_1^{\infty}$, where

$$a_k = \begin{cases} \frac{1}{k} & \text{if } n_k \\ 0 & \text{otherwise} \end{cases} \qquad \text{for the integers } n_1 < n_2 < \ldots,$$

is $(1,1)$-bounded multiplier.

Therefore we have the following:

Corollary 4.42. (Generalized Shields's inequality) *There is a constant $C > 1$ such that given any set $n_1 < n_2 < \ldots < n_k \subset \mathbb{Z}$, and $A = \sum_{k=1}^{\infty} A_{n_k} \in C_1$, we have that*

$$\sum_{k=1}^{\infty} \frac{||A_{n_k}||_{C_1}}{k} \leq C||A||_{C_1}.$$

Also Corollary 4.42 implies the matrix version of Corollary 4.39, which is the positive answer to a Littlewood conjecture.

Corollary 4.43. *There is a constant $C > 1$ such that, given any set $\{n_1 < n_2 < \ldots < n_N\} \subset \mathbb{Z}$ and matrix $A = \sum_{k=1}^{N} A_{n_k}$ with $||A_{n_k}||_{C_1} \geq 1$ for all k,*

$$||A||_{C_1} \geq C \log N.$$

Notes

The results of the first section were communicated to us by V. Lie. Proposition 4.4 is a Hardy-Littlewood type inequality and will be used in Chapter 8 to prove Theorem 8.6 which is the matrix version of a celebrated statement of M. Mateljevic and M.Pavlovic 1990 (see [60]).

The space $H^p(\ell_2)$ introduced in Definition 4.3 seems to be the right matrix version of classical Hardy space and can be used in many problems about matrix spaces suggested by the classical results in Hardy space theory and, more generally, in Harmonic Analysis.

Starting from a very interesting inequality of A. Shields [84] in Section 4.2 we get some new inequalities for the Pisier and Lust-Picquard spaces $T_p(\ell_R^2) + T_p(\ell_C^2)$ and $T_q(\ell_R^2) + T_q(\ell_C^2)$. It remains an open question if these inequalities may be replaced by more simple and more interesting matrix version of the Hardy-Littlewood inequality

$$\sum_{n \geq 0} (n+1)^{p-2}||A_n||_{C_p}^p \leq K(p)||A||_{T_p}^p,$$

for all upper triangular matrices $A \in C_p$.

The inequality of Shields is proved in Section 4.3 following the original proof from 1983.

Another more general proof of this inequality, due to O. Blasco and A. Pelczynski [15], is given in Section 4.4. Here the spaces of $(H^1 - \ell_1)$-Fourier type are studied. We present also the proof due to W. T. Sledd and D. A. Stegenga [88] of Ch. Fefferman's theorem about the $(H^1 - \ell_1)$ multipliers.

Also the proof of Sarason's noncommutative factorization theorem with the proof of U. Haagerup and G. Pisier is given. We mention the interesting and useful Parott's result about the completion of a Hankel matrix with $B(\ell_2)$-coefficients.

In an excellent monograph [68] dedicated to analytic functions on the disk M. Pavlovic characterized the analytic functions from H^1 by the condition

$$\sup_n \frac{1}{a_n} \sum_{j=0}^{n} ||s_j f|| < \infty.$$

Inspired by this result we proved in [73] a characterization of upper triangular matrices from C_1. This characterization is stated and proved in Section 4.5.

Finally, in Section 4.6 we present the proof of an extension of Shields's inequality.

Chapter 5

The matrix version of $BMOA$

5.1 First properties of $BMOA(\ell_2)$ space

We introduce different definitions for the matrix version of $BMOA$, which may be different from each other but on the class of Toeplitz matrices, they all coincide with the classical $BMOA$.

One of these definitions for the matrix version of the space $BMOA$, denoted for short by $BMOA(\ell_2)$, is as follows:

For $\lambda \in D$ we put $k_\lambda(e^{2\pi i\theta}) := \frac{1}{1-\bar{\lambda}e^{2\pi i\theta}}$ and denote by K_λ the Toeplitz matrix associated with the function $k_\lambda(\cdot)$.

Definition 5.1. *The space* $BMOA(\ell_2)$ *is defined by*

$$BMOA(\ell_2) := \{A \mid A \text{ upper triangular } ||A||_{BMOA(\ell_2)} < \infty\},$$

where

$$||A||^2_{BMOA(\ell_2)} := \sup_{\lambda \in D}\{\int_0^1 ||A'(r)K_{\bar{\lambda}r}||^2_{\mathcal{L}^2(\ell_2)}(1-r^2)(1-|\lambda|^2)r\,dr\}.$$

Here $\mathcal{L}^2(\ell_2)$ was introduced in Section 4.1.

We remark that $BMOA(\ell_2)$ is a Banach space endowed with the norm

$$||A_0||_{B(\ell_2)} + ||A||_{BMOA(\ell_2)}.$$

Our $BMOA(\ell_2)$ is a proper subspace of $BMOA_{\mathcal{C}}(\mathcal{L}^2(\ell_2))$ introduced by Blasco in [13].

Let $VMOA(\ell_2)$ be the subspace of $BMOA(\ell_2)$ consisting of those matrices A such that

$$\lim_{\lambda \to 1}\{\int_0^1 ||A'(r)K_{\bar{\lambda}r}||^2_{\mathcal{L}^2(\ell_2)}(1-r^2)(1-|\lambda|^2)r\,dr\} = 0.$$

We need the above definition of $BMOA(\ell_2)$ and the following results in order to extend a nice theorem of Mateljevic and Pavlovic (see [60]).

Proposition 5.2. *If $A \in \mathcal{L}_r^2(\ell_2)$, and $B \in B(\ell_2)$, then we have that*
$$||AB||_{\mathcal{L}_r^2(\ell_2)} \leq ||A||_{\mathcal{L}_r^2(\ell_2)}||B||_{B(\ell_2)}.$$

Proof. We note that
$$\mathcal{L}_{AB}(x,t) = \int_0^1 \mathcal{L}_B(x,s)\mathcal{L}_A(-s,t)ds,$$
which implies that
$$\mathcal{L}_{AB}(x,t) = \sum_{k=1}^{\infty} \mathcal{L}_k^B(x)\mathcal{C}_k^A(t) = \sum_{j=1}^{\infty} \left(\sum_{k=1}^{\infty} \mathcal{L}_k^B(x)a_{jk} \right) e^{2\pi ijt},$$
and, hence,
$$\mathcal{L}_j^{AB}(x) = \sum_{k=1}^{\infty} \mathcal{L}_k^B(x)a_{jk},$$
for all $j \geq 1$. Of course, here $A = (a_{jk})_{j\geq 1;\, k\geq 1}$, and $B = (b_{jk})_{j\geq 1;\, k\geq 1}$.

We deduce that
$$||AB||_{\mathcal{L}_r^2(\ell_2)} = \sup_{j\geq 1} \left(\int_0^1 |\mathcal{L}_j^{AB}(x)|^2 dx \right)^{1/2} =$$
$$\sup_{j\geq 1} \sup_{||h||_{L^2}\leq 1} | \int_0^1 \left(\sum_{k=1}^{\infty} \mathcal{L}_k^B(x)a_{jk} \right) h(x)dx | \leq$$
$$\sup_{j\geq 1} \sup_{||h||_{L^2}\leq 1} \left(\sum_{k=1}^{\infty} |a_{jk}|^2 \right)^{1/2} \left(\sum_{k=1}^{\infty} | \int_0^1 \mathcal{L}_k^B(x)h(x)dx |^2 \right)^{1/2}$$
$$= ||A||_{\mathcal{L}_r^2(\ell_2)}||B||_{B(\ell_2)}$$
and the proof is complete. □

Remark 5.3. *By reasoning as in the proof of Proposition 5.2 we can derive that*
$$||AB||_{\mathcal{L}_r^2(\ell_2)} \leq ||A||_{B(\ell_2)}||B||_{\mathcal{L}_c^2(\ell_2)}. \tag{5.1}$$
If the matrix B is a Toeplitz matrix, then
$$||B||_{\mathcal{L}_r^2(\ell_2)} = ||B||_{\mathcal{L}_c^2(\ell_2)}$$
and, consequently, (5.1) is equivalent to that
$$||AB||_{\mathcal{L}_r^2(\ell_2)} \leq ||A||_{B(\ell_2)}||B||_{\mathcal{L}_r^2(\ell_2)}$$
for B a Toeplitz matrix, and we may delete the subscript r in the above formulas.

We give now a sufficient condition to guarantee that an upper triangular matrix belongs to $BMOA(\ell_2)$.

Corollary 5.4. *If A is a matrix of finite band type $A = \sum_{k=0}^{n} A_k$, then we have that*

$$\|A\|_{BMOA(\ell_2)} \leq \left(\int_0^1 \|A'(r)\|_{B(\ell_2)}^2 (1 - r)dr \right)^{1/2}.$$

Proof.

$$\|A\|_{BMOA(\ell_2)} \leq \text{(by Proposition 5.2 and Remark 5.3)} \leq$$

$$C \sup_{\lambda \in (0,1)} \left\{ \int_0^1 \|A'(r)\|_{B(\ell_2)}^2 \|K_{\overline{\lambda}r}\|_{\mathcal{L}^2(\ell_2)}^2 (1 - r)(1 - |\lambda|^2)rdr \right\}^{1/2}.$$

We note that

$$\|K_{\overline{\lambda}r}\|_{\mathcal{L}^2(\ell_2)}^2 = \int_0^1 \frac{d\theta}{|1 - \overline{\lambda}re^{2\pi i\theta}|^2} = \frac{1}{1 - |\lambda r|^2},$$

and since $\frac{(1-|\lambda|^2)r}{1-|\lambda r|^2} \leq 1$ we are done. $\square$

See also Proposition 1.2-[13] where a rather strong result was proved for the Blasco space $BMOA_C(X)$ with an arbitrary Banach space X.

Inspired by Proposition 1.3-[13] we give an example of a matrix belonging to $BMOA(\ell_2)$.

Let us denote first by e_{ij}, for fixed $i \geq 1$, $j \geq 1$, the matrix having 1 on the intersection of the ith row with the jth column and 0 otherwise.

Example 5.5. *Let $A = \sum_{k=1}^{\infty} \frac{1}{n \ln(n+1)} e_{n,2n}$. Then the computations done in [13]-Proposition 1.3 give us that*

$$\|A\|_{BMOA(\ell_2)} \leq C \int_0^1 \frac{dr}{(1 - r)(\ln(\frac{1}{1-r}))^2} < \infty.$$

5.2 Another matrix version of *BMO* and matriceal Hankel operators

In the last thirty years a number of important research papers are devoted to Hankel operators. See for instance the very impressive monograph [65].

In that book there is a chapter dedicated to vectorial Hankel operators, which are important for applications and now we recall some notions and definitions from this chapter.

Let us denote by $H^2(\mathcal{H})$ the Hardy class of functions with values in a separable Hilbert space $\mathcal{H}$ and by $B(\mathcal{H}, \mathcal{K})$ the space of bounded linear operators from $\mathcal{H}$ to $\mathcal{K}$ and by $L^\infty(B(\mathcal{H}, \mathcal{K}))$ the space of bounded weakly measurable $B(\mathcal{H}, \mathcal{K})$-valued functions. We recall that the Hardy class $H^2(\mathcal{H})$ is defined as follows:

$$H^2(\mathcal{H}) = \{F \in L^2(\mathcal{H}) : \hat{F}(n) = 0, \ n < 0\},$$

where $L^2(\mathcal{H})$ is the space of weakly measurable $\mathcal{H}$-valued functions F for which

$$||F||^2_{L^2(\mathcal{H})} = \int_{\mathbb{T}} ||F(\zeta)||^2_{\mathcal{H}} dm(\zeta) < \infty.$$

Put $H^2_-(\mathcal{H}) = L^2(\mathcal{H}) \ominus H^2(\mathcal{H})$. Let $\mathcal{K}$ be another separable Hilbert space and let Φ be a function in $L^2_s(B(\mathcal{H}, \mathcal{K}))$ (see [65], chapter 2), i.e.

$$\int_{\mathbb{T}} ||\Phi(\zeta)x||^2_{\mathcal{K}} dm(\zeta) < \infty \text{ for any } x \in \mathcal{H}. \tag{5.2}$$

Recall that for functions satisfying (5.2) the Fourier coefficients $\hat{\Phi}(n) \in B(\mathcal{H}, \mathcal{K})$ are defined by

$$\hat{\Phi}(n)x = \int_{\mathbb{T}} \bar{\zeta}^n \Phi(\zeta)x dm(\zeta), n \in \mathbb{Z}, \ x \in \mathcal{H}.$$

Now we can define for such functions Φ the Hankel operator $H_\Phi : H^2(\mathcal{H}) \to H^2_-(\mathcal{K})$ on the set of polynomials in $H^2(\mathcal{H})$ by

$$H_\Phi F = \mathbb{P}_- \Phi F, \ F \in H^2(\mathcal{H}), \tag{5.3}$$

where we denote by $\mathbb{P}_+$ and $\mathbb{P}_-$ the orthogonal projections onto $H^2(\mathcal{H})$ and $H^2_-(\mathcal{K})$, respectively.

We denote by C_2 the set of all Hilbert-Schmidt matrices, equipped with the usual Hilbert-Schmidt norm $|| \cdot ||_{C_2}$.

Let T_2 be the set of all upper triangular Hilbert-Schmidt matrices endowed with the norm induced by $|| \cdot ||_{C_2}$. Of course T_2 is a Hilbert space.

If $A \in T_2$ and $k = 0, 1, 2, 3, \dots$ we denote by A_k the kth-diagonal of A. Then, formally, $A = \sum_{k \in \mathbb{Z}} A_k$ and we denote by P_+ the triangular projection $P_+ A = \sum_{k \geq 0} A_k$. It is well-known that $P_+ : T_2 \to T_2$ is bounded with norm less than 1.

Now put $P_- = I - P_+$, where I is the identity on T_2. Of course $||P_-|| \leq 1$. Put

$$H^\infty(B(T_2, T_2)) := \{\Phi \in L^\infty(B(T_2, T_2)) : \hat{\Phi}(n) = 0, n < 0\}.$$

Then the following statement is known (see [65]):

Vectorial Nehari Theorem. *Let Φ be a $B(T_2, T_2)$-valued function that satifies (5.2). Then the operator H_Φ defined by (5.3) extends to a bounded operator on $H^2(\mathcal{H})$ if and only if there exists a function Ψ in $L^\infty(B(T_2, T_2))$ such that*

$$\hat{\Psi}(n) = \hat{\Phi}(n), \ n < 0,$$

and

$$\|H_\Phi\| = dist_{L^\infty}(\Psi, H^\infty(B(T_2, T_2))).$$

Let Φ be an infinite matrix such that for all matrices $A = \sum_{k=0}^{n} A_k$, $n \in \mathbb{N}$, we have that $P_-(\Phi A) \in C_2$. Examples of such matrices are either all matrices representing linear bounded operators on ℓ_2, or matrices Φ such that $P_-\Phi = 0$.

We define the *matrix version of Hankel operator H_Φ* as follows: $H_\Phi : T_2 \to (T_2)_- := C_2 \ominus T_2$, on the dense subspace in T_2 of all matrices $A = \sum_{k=0}^{n} A_k$, $n \in \mathbb{N}$, such that $\Phi A \in C_2$, by

$$H_\Phi(A) = P_-(\Phi A).$$

The matrix Φ is called *the symbol of the Hankel operator H_Φ*.

If Φ is an infinite matrix and if we denote by Φ also the operator function given by $\Phi(\zeta) = \sum_{k \in \mathbb{Z}} e^{ik\zeta} \Phi_k$, then it is easy to verify that the matrix version of the Hankel operator H_Φ defined on T_2 coincides with the restriction of Peller's vectorial Hankel operator H_Φ to the subspace of $H^2(T_2)$ consisting of all matrix functions $\sum_{k=0}^{\infty} e^{ik\zeta} A_k$, for some $A \in T_2$.

In 1985 S. Power [77] gave the matrix versions of Nehari's and Hartman's theorems.

In what follows we present these results about matrix versions of Hankel operators H_Φ with different proofs.

We denote as in [84], by $\tilde{A}$ the matrix whose entries $\tilde{a}_{kl}$ are given by

$$\tilde{a}_{kl} = \begin{cases} -ia_{kl} & \text{if } k < l, \\ ia_{kl} & \text{if } k > l, \\ 0 & \text{if } k = l, \end{cases}$$

where a_{kl} are the entries of A.

We recall that $B(\ell_2)$ means the space of representing matrices of all linear bounded operators on ℓ_2, equipped with the usual operator norm.

Moreover, we denote by $BMO_F(\ell_2)$ the space of all matrices A such that there exist matrices Φ and Ψ from $B(\ell_2)$ with $A = \Phi + \tilde{\Psi}$, equipped with the norm $||A||_{BMO_F(\ell_2)} := \inf\{||\Phi||_{B(\ell_2)} + ||\Psi||_{B(\ell_2)}; A = \Phi + \tilde{\Psi}\}$. Of course $BMO_F(\ell_2)$ is a Banach space. We also define $BMOA_F := BMO_F(\ell_2) \cap H^\infty(\ell_2)$, where $H^\infty(\ell_2)$ is the set of all upper triangular matrices $A \in B(\ell_2)$.

Then the following holds [69] (for a more general result see [77], [59]):

Theorem 5.6. (Matrix version of the Nehari theorem) *Let Φ be an infinite matrix such that $\Phi A \in C_2$ for all finite band-type matrices A.*

Then the following conditions are equivalent:

(a) H_Φ is a bounded linear operator on T_2.

(b) There is $\Psi \in B(\ell_2)$ such that $\Psi_k = \Phi_k$ for all $k < 0$.

(c) $P_-\Phi \in BMO_F(\ell_2)$.

Proof.   (b)$\Rightarrow$(a). By simple computations we get that $H_\Phi(A) = H_\Psi(A)$ for all $A = \sum_{k=0}^n A_k$, $n \in \mathbb{N}$. Hence,

$$||H_\Phi(A)||_{C_2} = ||H_\Psi(A)||_{C_2} \le ||\Psi A||_{C_2} \le ||\Psi||_{B(\ell_2)}||A||_{T_2},$$

that is

$$||H_\Phi|| \le ||\Psi||_{B(\ell_2)} < \infty.$$

(a)$\Rightarrow$ (b). We note that $H_\Phi(A) = P_-(\Phi A) = P_-[P_-(\Phi)A]$ for all $A = \sum_{k=0}^n A_k$, $n \in \mathbb{N}$. Assume that $H_\Phi : T_2 \to (T_2)_-$ is a bounded linear operator. Then

$$||H_\Phi(A)||_{(T_2)_-} = \sup_{||B||_{(T_2)_-} \le 1} | < H_\Phi(A), B > | = \sup_{||B||_{(T_2)_-} \le 1} |tr\, H_\Phi(A)B^*| =$$

$$\sup_{||B||_{(T_2)_-} \le 1} |tr\, P_-[(P_-\Phi)A]B^*|. \tag{5.4}$$

But, obviously $tr\, P_-[(P_-\Psi)A]B^* = tr\, (P_-\Phi)AB^*$ for $B \in (T_2)_-$ and $A \in T_2$.

Hence, (5.4) can be written as

$$||H_\Phi(A)||_{(T_2)_-} = \sup_{||B||_{(T_2)_-} \le 1} |tr\, (P_-\Phi)(AB^*)|.$$

Since $A, B^* \in T_2$ and every $A \in T_1$ may be written as $A = B_1 B_2$, where $B_1, B_2 \in T_2$ and $||B_1||_{T_2} = ||B_2||_{T_2} = ||A||_{T_1}^{1/2}$ (see [84]), we have that

$$\infty > M = ||H_\Phi|| = \sup_{||A||_{(T_2)} \le 1, ||B||_{(T_2)_-} \le 1} |tr\, (P_-\Phi)AB^*| =$$

$$\sup_{||A||_{(T_1)} \leq 1;\ A_0 = 0} |tr\,(P_-\Phi)A| = ||P_-\Phi||_{(T_{1,0})^*},$$

where $T_{1,0} = \{A \in T_1; A = \sum_{j \geq 1} A_j\}$. Using the Hahn-Banach theorem we get that

$$\infty > ||H_\Phi|| = \inf_{P_-\Phi - \Psi \in (T_{1,0})^\perp} ||\Psi||_{B(\ell_2)}.$$

But

$$(T_{1,0})^\perp = \{\Psi \in B(\ell_2)|\, tr(\Psi B) = 0 \text{ for all } B \in T_{1,0}\} =$$

$$\{\Psi \in B(\ell_2)|\, P_-\Psi = 0\}$$

and, hence, there is $\Psi \in B(\ell_2)$ such that $P_-\Phi = P_-\Psi$. Thus (b) holds.

(b) $\Rightarrow$ (c). $P_-\Phi = P_-\Psi$ for $\Psi \in B(\ell_2)$. But

$$P_-\Psi = -i\tilde{\Psi} + P_+\Psi - \Psi_0 = -i\tilde{\Psi} + \Psi - P_-\Psi - \Psi_0$$

and, thus, $P_-\Psi = -\frac{1}{2}i\tilde{\Psi} + \frac{1}{2}\Psi - \frac{1}{2}\Psi_0$, that is $P_-\Psi \in BMO_F(\ell_2)$.

(c) $\Rightarrow$ (b). Let $P_-\Phi \in BMO_F(\ell_2)$. Then there exist $A, B \in B(\ell_2)$ with $P_-\Phi = A + \tilde{B}$. But

$$P_-\Phi = P_-(A) + P_-(\tilde{B}) = P_-(A) + iP_-B = P_-(A + iB) = P_-\Psi,\ \Psi \in B(\ell_2),$$

and the implication follows. The proof is complete. $\qquad\square$

Corollary 5.7. *It yields that*

$$||H_\Phi|| = \inf\{||\Phi - F||_{B(\ell_2)}; F \in H^\infty(\ell_2)\} = dist_{B(\ell_2)}(\Phi, H^\infty(\ell_2)).$$

Proof. It follows directly from the equality

$$||H_\Phi|| = \inf\{||\Psi||_{B(\ell_2)}; \Psi - P_-\Phi \in (T_{1,0})^\perp\}.$$

$\qquad\square$

We intend to give a matrix version of Hartman's Theorem (see [65]) about compacity of Hankel operators.

We need the following lemma:

Lemma 5.8. *Let $\Phi \in C_\infty$, the space of all compact operators on ℓ_2. Then $dist_{B(\ell_2)}(\Phi, H^\infty(\ell_2)) = dist_{B(\ell_2)}(\Phi, T_\infty)$, where $T_\infty = \{A \in C_\infty |\, A \text{ upper triangular}\}$.*

Proof. The inequality $dist_{B(\ell_2)}(\Phi, H^\infty(\ell_2)) \leq dist_{B(\ell_2)}(\Phi, T_\infty)$ is trivial. Conversely, let $\Phi \in C_\infty$, $H \in H^\infty(\ell_2)$ and P_n be the projection on $B(\ell_2)$, given by $P_n A = (a'_{ij})_{i,j\geq 1}$, where

$$a'_{ij} = \begin{cases} a_{ij} & \text{if } 1 \leq 1, j \leq n \\ 0 & \text{otherwise} \end{cases}, \quad n = 1, 2, 3, \ldots.$$

Then, since Φ is a compact operator, for each $\epsilon > 0$ there exists $n_0 = n_0(\epsilon)$ such that $||\Phi - P_{n_0}\Phi|| < \epsilon$. Therefore,

$$||\Phi - H||_{B(\ell_2)} \geq ||P_{n_0}\Phi - P_{n_0}H||_{B(\ell_2)} \geq ||\Phi - P_{n_0}H||_{B(\ell_2)} - \epsilon \geq$$
$$dist_{B(\ell_2)}(\Phi, T_\infty) - \epsilon$$

and, since ϵ is arbitrary, we have that

$$dist_{B(\ell_2)}(\Phi, H^\infty(\ell_2)) \geq dist_{B(\ell_2)}(\Phi, T_\infty).$$

The proof is complete. $\qquad\square$

Theorem 5.9. $H^\infty(\ell_2) + C_\infty$ *is a closed Schur subalgebra of* $B(\ell_2)$.

Proof. Lemma 5.8 shows that the canonical inclusion $C_\infty/T_\infty \hookrightarrow B(\ell_2)/H^\infty(\ell_2)$ is an isometry and, therefore, C_∞/T_∞ is a closed subspace of $B(\ell_2)/H^\infty(\ell_2)$.

Let $\rho : B(\ell_2) \to B(\ell_2)/H^\infty(\ell_2)$ be the canonical map. It follows that $H^\infty(\ell_2) + C_\infty = \rho^{-1}(C_\infty/T_\infty)$ is a closed subspace of $B(\ell_2)$.

Now we should show that $H^\infty(\ell_2) + C_\infty$ is a Schur subalgebra of $B(\ell_2)$. But, in fact,

$$(A_1 + B_1) * (A_2 + B_2) = (A_1 * A_2 + B_1 * A_2 + A_1 * B_2) + B_1 * B_2,$$

for all $A_1, A_2 \in H^\infty(\ell_2)$ and $B_1, B_2 \in C_\infty$.

The first expression on the right hand side of the above equality belongs to $H^\infty(\ell_2)$ by [11]. On the other hand $B_1 * B_2 \in C_\infty$, since $\lim_n P_n B_i = B_i$, $i = 1, 2$, again by [11]. $\qquad\square$

Now let us denote by S the linear and bounded operator on C_2 given by $S(A) = \begin{pmatrix} a^1{}_1 & a^1_0 & a^1_1 & \cdots \\ a^1_{-2} & a^2_{-1} & a^2_0 & \ddots \\ a^1_{-3} & \ddots & \ddots & \ddots \\ \vdots & \ddots & \ddots & \ddots \end{pmatrix}$ for $A = \begin{pmatrix} a^1_0 & a^1_1 & a^1_2 & \cdots \\ a^1_{-1} & a^2_0 & a^2_1 & \ddots \\ a^1_{-2} & a^2_{-1} & \ddots & \ddots \\ \vdots & \ddots & \ddots & \ddots \end{pmatrix}$.

Then we have the following:

Lemma 5.10. *Let* $K : T_2 \to (T_2)_-$ *be a compact operator and* $S : T_2 \to T_2$. *Then*

$$\lim_{n\to\infty} ||KS^n|| = 0.$$

Proof. It is enough to prove the lemma for rank one operators K.

Let $K(A) = <A, B> C$, where $B \in T_2$, $C \in (T_2)_-$. Then we have that $K S^n(A) = <S^n(A), B> C = <A, S^{*n}(B)> C$, where

$$S^*(A) = \begin{pmatrix} a_1^1 & a_2^1 & a_3^1 & \cdots \\ 0 & a_1^2 & a_2^2 & \ddots \\ 0 & 0 & \ddots & \ddots \\ \vdots & \ddots & \ddots & \ddots \end{pmatrix} \quad \text{for } A = \begin{pmatrix} a_0^1 & a_1^1 & a_2^1 & \cdots \\ 0 & a_0^2 & a_1^2 & \ddots \\ 0 & 0 & \ddots & \ddots \\ \vdots & \ddots & \ddots & \ddots \end{pmatrix} \in T_2.$$

Hence $\|KS^n\| = \|S^{*n}B\|_{T_2} \|C\|_{T_2} \to 0$. $\qquad\square$

Lemma 5.11. *Let $\Phi \in C_\infty$. Then $H_\Phi : T_2 \to (T_2)_-$ is a compact operator.*

Proof. Let $F = e_{ij}$, for fixed $i, j \geq 1$ such that $i - j \geq 1$. Then $H_F(A) = \sum_{k=0}^{i-j-1} a_k^j e_{i,k+1}$, where A has the entries $(a_k^j)_{j=1}^\infty$ on the kth-diagonal, for $k \geq 0$.

Therefore H_F is a compact operator and, consequently, it is easy to see that $H_{P_n \Phi}$ is a compact operator too, for all $n \geq 0$.

By Theorem 5.6, and using the fact that Φ is a compact operator, we have that

$$\|H_\Phi - H_{P_n \Phi}\| \leq \|\Phi - P_n \Phi\|_{B(\ell_2)} \to 0.$$

Consequently, H_Φ is a compact operator. $\qquad\square$

Now we are ready to state the matrix version of the Hartman theorem (see [69], [77]):

Theorem 5.12. *Let $\Phi \in B(\ell_2)$. The following conditions are equivalent:*
(i) H_Φ is a compact operator.
(ii) $\Phi \in H^\infty(\ell_2) + C_\infty$.
(iii) There exists $\Psi \in C_\infty$ with $H_\Phi = H_\Psi$.

Proof. Obviously (ii)$\Rightarrow$ (iii). Since (iii) implies that $\Phi_k = \Psi_k$ for all $k < 0$, the assertion (i) follows in the same way as (b)$\Rightarrow$ (a) in Theorem 5.6.

(i) $\Rightarrow$ (ii). We have that

$$\|H_\Phi S^n\| = \|H_{S^n \Phi}\| = \text{(by Corollary 5.7)} = dist_{B(\ell_2)}(\Phi, (S^{-1})^n H^\infty(\ell_2)).$$

Then, for all $\epsilon > 0$ and for all $n \geq 0$ there exist $B^n \in H^\infty(\ell_2)$ and a matrix A^n of finite band type such that $A^n = \sum_{j=-1}^{-n} (A^n)_j$; $A^n \in B(\ell_2)$ and, by using Theorem 5.6,

$$\|H_\Phi S^n\|_{B(\ell_2)} \geq \|\Phi - A^n - B^n\|_{B(\ell_2)} - \epsilon \geq \|H_\Phi - H_{A^n}\| - \epsilon$$

$$\geq ||P_m H_\Phi - P_m H_{A^n}|| - \epsilon \text{ for all } m \geq 1.$$

Hence,

$$||H_\Phi S^n|| \geq ||H_\Phi - P_m H_{A^n}|| - ||H_\Phi - P_m H_\Phi|| - \epsilon \text{ for all } m, n.$$

But $P_m H_{A^n} = H_{P_m(A^n)}$, and, therefore,

$$||H_\Phi S^n|| \geq ||H_\Phi - H_{P_m(A^n)}|| - ||H_\Phi - P_m H_\Phi|| - \epsilon \quad \forall m, n.$$

Moreover, by hypothesis, H_Φ is a compact operator, and, consequently, there exist m_0 and $C \in H^\infty(\ell_2)$ such that $||H_\Phi - P_{m_0} H_\Phi|| < \epsilon$, and

$$||H_\Phi S^n|| \geq ||H_\Phi - H_{P_m(A^n)}|| - 2\epsilon$$

$$\geq \text{(by Corollary 5.7)} \geq ||\Phi - P_{m_0}(A^n) - C||_{B(\ell_2)} - 3\epsilon, \text{ for all } n.$$

By Lemma 5.10 we get that

$$\epsilon \geq dist_{B(\ell_2)}(\Phi, H^\infty(\ell_2) + C_\infty) - 3\epsilon.$$

Therefore $dist_{B(\ell_2)}(\Phi, H^\infty(\ell_2) + C_\infty) = 0$ and, by Theorem 5.9, it follows that $\Phi \in H^\infty(\ell_2) + C_\infty$. The proof is complete.   □

If we denote by

$$VMO_F(\ell_2) := \{\Phi | \Phi = A + \tilde{B}, A, B \in C_\infty\}$$

equipped with the norm induced by that of $BMO_F(\ell_2)$, then, as in the proof (b) $\Rightarrow$ (c) of Theorem 5.6, we can derive the following statement:

Theorem 5.13. *Let Φ be an infinite matrix. Then H_Φ is a compact operator if and only if $P_-\Phi \in VMO_F(\ell_2)$.*

$BMOA_F(\ell_2)$ is the space of all upper triangular matrices $A \in BMO_F(\ell_2)$.

This space is of interest because the proof of Theorem 5.6- (b) $\Rightarrow$ (c), Hahn-Banach theorem and the well-known fact [34] that $C_1^* = B(\ell_2)$ via the bilinear map

$$< A, B > = \text{tr } AB^* \quad \text{for } A \in B(\ell_2) \text{ and } B \in C_1$$

show us that

$$T_1^* = BMOA_F(\ell_2)$$

by the previous bilinear map. See also Theorem 2.3 [77].

5.3 Nuclear Hankel operators and the space $\mathcal{M}^{1,2}$

We shall here derive a sufficient condition to guarantee that the Hankel operator H_A introduced in Section 5.2, where A is an upper triangular matrix, is nuclear. Moreover, if the matrix A has only a finite number of rows this condition is also necessary.

Let $1 \leq p < \infty$. We denote by S_p the Schatten class of all operators from T_2 into T_{2-}.

We denote by $\mathcal{M}^{p,2}$ the space of all upper triangular infinite matrices $A = (a_k^l)_{k\geq 0, l\geq 1}$ such that

$$||A||_{p,2} := \left(\sum_{l=1}^{\infty} \sum_{j=0}^{\infty} \left(\sum_{n=j}^{\infty} |a_n^l|^2 \right)^{p/2} \right)^{1/p} < \infty. \qquad (5.5)$$

Our next result reads:

Theorem 5.14. *Let A be an upper triangular infinite matrix such that $A \in \mathcal{M}^{1,2}$. Then the Hankel operator $H_{A^*} : T_2 \to T_{2-}$ is nuclear and, moreover,*

$$||H_{A^*}||_{S_1} \leq ||A||_{\mathcal{M}^{1,2}}.$$

Proof. We easily see that

$$H_{A^*}(B) = \sum_{l=0}^{\infty} \sum_{k=1}^{\infty} \left(\sum_{j=l+1}^{\infty} |a_j^k|^2 \right)^{1/2} \cdot < E_l^k, B > A^{kl}, \qquad (5.6)$$

where $B \in T_2$, $A^{kl} = \left(\sum_{j=l+1}^{\infty} \overline{a_j^k} E_{-j+l}^{k+l} \right) \left(\sum_{j=l+1}^{\infty} |a_j^k|^2 \right)^{-1/2}$, the series (5.6) converges in the norm of C_2 and E_j^k, where $j \in \mathbb{Z}$, and $k \geq 1$, is the infinite matrix having as entries 1 on the kth place on the jth-diagonal, and 0 otherwise.

Consequently,

$$H_{A^*} = \sum_{l=0}^{\infty} \sum_{k=1}^{\infty} \left(\sum_{j=l+1}^{\infty} |a_j^k|^2 \right)^{1/2} \cdot E_l^k \otimes A^{kl}$$

is a nuclear operator for $A = A \in \mathcal{M}^{1,2}$ and $||H_{A^*}||_{S_1} \leq ||A||_{\mathcal{M}^{1,2}}$. The proof is complete. $\qquad \square$

Notes

In Section 5.1 we introduce the matrix versions of the classical spaces BMO and $BMOA$, $BMO(\ell_2)$ and $BMOA(\ell_2)$. These spaces are proper subspaces of that introduced previously by O. Blasco in [13]. These definitions were invented also by V. Lie and he kindly communicated his results to us. We also mention a kind of Hölder's inequality (Proposition 5.2). These results will be used in Chapter 8.

Another kind of matrix version of a BMO space is introduced in Section 5.2. This space, denoted by $BMO_F(\ell_2)$, is useful in the study of the matrix version of Hankel operator H_Φ and in a more general context of nest algebras it was introduced by S. Power in [77].

Theorem 5.14 seems to be new. It is also easy to observe that a converse of this theorem holds for matrices A with a finite number of rows.

Chapter 6

Matrix version of Bergman spaces

6.1 Schatten class version of Bergman spaces

We intend now to introduce the Schatten class version of Bergman spaces. In order to do this we apply Proposition 1.1 to obtain that, for a w^*-measurable $B(\ell_2)$-valued function f, the function $t \to ||f(t)||_{B(\ell_2)}$ is Lebesgue measurable on D. We introduce also the following matrix spaces:

$$L^\infty(D, \ell_2) := \{r \to A(r) \text{ being a } w^*\text{-measurable function on } [0,1) \,|$$

$$\text{ess} \sup_{0 \leq r < 1} ||A(r)||_{B(\ell_2)} := ||A(r)||_{L^\infty(D, \ell_2)} < \infty\},$$

$\widetilde{L^\infty}(D, \ell_2)$, is the subspace of $L^\infty(D, \ell_2)$ consisting of all strong measurable functions on $[0,1)$ and

$$L_a^\infty(D, \ell_2) := \{A \text{ infinite analytic matrix} \,|\, ||A||_{L_a^\infty(D,\ell_2)} :=$$

$$\sup_{0 \leq r < 1} ||C(r) * A||_{B(\ell_2)} = ||A(r)||_{L^\infty(D,\ell_2)} < \infty\}.$$

Definition 6.1. Let $1 \leq p < \infty$. We denote by $L^p(D, \ell_2)$ the space of all strong measurable C_p-valued functions defined on $[0,1)$, such that $||A(r)||_{L^p(D,\ell_2)} := \left(2 \int_0^1 ||A(r)||_{C_p}^p \, r \, dr\right)^{1/p} < \infty$, where C_p is the Schatten class of order p, and we define $\tilde{L}_a^p(D, \ell_2)$ as the space of all functions $A(r) := A * C(r)$, where A is an upper triangular matrix with $||A||_{L^p(D,\ell_2)} < \infty$.

Here of course $C(r)$ means the Toeplitz matrix associated to the Cauchy kernel $\frac{1}{1-r}$, for $0 \leq r < 1$ and $*$ stands for Schur product.

$\tilde{L}_a^p(D, \ell_2)$ is a subspace of $L^p(D, \ell_2)$.

By $L_a^p(D, \ell_2)$ we mean the space of all upper triangular matrices such that $||A(\cdot)||_{L^p(D,\ell_2)} < \infty$.

121

We identify $\tilde{L}_a^p(D, \ell_2)$ and $L_a^p(D, \ell_2)$ and call $L_a^p(D, \ell_2)$ the *Bergman-Schatten classes.*

We intend to introduce the concept of *Bergman projection* and we prove first some introductory results:

Lemma 6.2. *Let A be an upper triangular matrix belonging either to C_p, $1 \leq p \leq \infty$, or to $B(\ell_2)$. Then the function $r \to A(r)$ is a continuous function on $[0,1]$ taking values in C_p, $1 \leq p \leq \infty$, or in $B(\ell_2)$, too.*

Proof.   If $A \in B(\ell_2)$, then we have, when $r_n \to r \in [0,1]$, that

$$\|(C(r_n) - C(r)) * A\|_{B(\ell_2)} \leq \text{(by Theorem 2.14)}$$

$$\leq \frac{|r_n - r|}{|1 - r_n| \, |1 - r|} \|A\|_{B(\ell_2)} \underset{n \to \infty}{\to} 0.$$

Using the duality and interpolation between C_p we have that

$$\lim_{n \to \infty} \|(C(r_n) - C(r)) * A\|_{C_p} = 0, \quad 1 \leq p \leq \infty,$$

if $A \in C_p$.   □

By Lemma 6.2 we have:

Corollary 6.3. *Let $1 \leq p \leq \infty$ and let A be an upper triangular matrix. If $A \in C_p$ (respectively if $A \in B(\ell_2)$), then*

$$\sup_{0 \leq r < 1} \|C(r) * A\|_{C_p} = \sup_{0 \leq r \leq 1} \|C(r) * A\|_{C_p}$$

and similarly with $\| \; \|_{B(\ell_2)}$ instead of $\| \; \|_{C_p}$.

Proof.   We have to show that, for $1 \leq p \leq \infty$,

$$\|C(1) * A\|_{C_p} \leq \sup_{0 \leq r < 1} \|C(r) * A\|_{C_p},$$

(respectively the similar inequality for $\| \; \|_{B(\ell_2)}$.)

According to Lemma 6.2 we have that

$$\|A\|_{C_p} = \|C(1) * A\|_{C_p} = \lim_{r \to 1} \|C(r) * A\|_{C_p} \leq \sup_{r < 1} \|C(r) * A\|_{C_p}$$

and similar for $B(\ell_2)$.   □

In the sequel we denote $L_a^\infty(D, \ell_2)$ by $H^\infty(D, \ell_2)$, or simply by $H^\infty(\ell_2)$.

Corollary 6.4. *$H^\infty(\ell_2)$ is a Banach subspace in $B(\ell_2)$.*

Proof. We have that

$$||A||_{L_a^\infty(D,\ell_2)} = \sup_{0 \le r < 1} ||C(r) * A||_{B(\ell_2)} = \text{(by Corollary 6.3)}$$

$$= \sup_{0 \le r \le 1} ||C(r) * A||_{B(\ell_2)}$$

$$= \sup_{r \in [0,1]} \sup_{||B||_{C_1} \le 1, B \text{ a lower triangular matrix}} |\text{tr}(C(r) * A)B|$$

$$\le ||A||_{B(\ell_2)} \sup_{||B||_{C_1} \le 1, \text{rank}(B) < \infty, B \text{ lower triangular}} ||C(r) * B||_{C_1}$$

$$\le \text{(by Lemma 6.2)} \le ||A||_{B(\ell_2)}.$$

On the other hand

$$||A||_{L_a^\infty(D,\ell_2)} = \sup_{0 \le r \le 1} ||C(r) * A||_{B(\ell_2)} \ge ||A||_{B(\ell_2)},$$

and, consequently, $||A||_{L_a^\infty(D,\ell_2)} \sim ||A||_{B(\ell_2)}$. $\qquad \square$

Proposition 6.5. *Let* $1 \le p < \infty$. *Then* $\tilde{L}_a^p(D,\ell_2)$ *is a closed subspace in* $L^p(D,\ell_2)$, *and, thus,* $L_a^p(D,\ell_2)$ *may be identified by the map* $A \to A(r)$, $r \in [0,1)$, *with a closed subspace of* $L^p(D,\ell_2)$. *Consequently, the Bergman-Schatten space* $L_a^p(D,\ell_2)$ *is a Banach space.*

Proof. Let $A^n \in L_a^p(D,\ell_2)$. If $A^n(r) \to A(r) \in L^p(D,\ell_2)$, then $(A^n(r))_n$ is a Cauchy sequence in $L^p(D,\ell_2)$ and, hence $||C(r) * (A^n - A^m)||_{C_p} \xrightarrow[n,m \to \infty]{} 0$ a.e. with respect to the Lebesgue measure on $[0,1]$. Consequently, by Lemma 6.2, it follows that $\lim_{n,m \to \infty} ||C(r) * (A^n - A^m)||_{C_p} = 0$ for all $r \in [0,1]$, that is the sequence $(C(r) * A^n)_n$ is a Cauchy sequence in C_p for all $r \in [0,1]$, which in turn implies that $\lim_{n \to \infty}(C(r)*A^n)(i,j) = A(r)(i,j)$ for all $i,j \in \mathbb{N}$ and for all $0 \le r \le 1$. We conclude that $A(r)$ is an upper triangular matrix for all $r \le 1$ and since $(A^n * C(r))(i,j) = a_{ij}^n r^{j-i}$ it follows that there is $\lim_{n \to \infty} a_{ij}^n = a_{ij}$ for all $i,j \in \mathbb{N}$ and $A(r) = C(r) * A$, where $A = (a_{ij})_{i,j}$. Thus $A \in L_a^p(D,\ell_2)$. $\qquad \square$

Remark. We note that in the above proof we have that $\lim_{r \to 1} ||A - A(r)||_{L_a^p(D,\ell_2)} = 0$. It is also clear that the matrices of finite-band type are dense in $L_a^p(D,\ell_2)$. See also [14]-Proposition 2.3 for related results.

There are some general results about Bergman-Schatten classes essentially dues to O. Blasco [14].

Proposition 6.6. *Let* $1 \leq p < \infty$ *and*

$$A = \begin{pmatrix} a_0^1 & a_1^1 & a_2^1 & \cdots \\ 0 & a_0^2 & a_1^2 & \ddots \\ 0 & 0 & a_0^3 & \ddots \\ \vdots & \vdots & \vdots & \ddots \end{pmatrix}$$

be an upper triangular infinite matrix Then $A \in L_a^p(D, \ell_2)$ *if and only if*

$$B := \begin{pmatrix} a_1^1 & a_2^1 & a_3^1 & \cdots \\ 0 & a_1^2 & a_2^2 & \ddots \\ 0 & 0 & a_1^3 & \ddots \\ \vdots & \vdots & \vdots & \ddots \end{pmatrix} \in L_a^p(D, \ell_2).$$

Moreover, there exists a positive constant $K < 1$ *such that*

$$K(\|A_0\|_{C_p} + \|B\|_{L_a^p(D,\ell_2)}) \leq \|A\|_{L_a^p(D,\ell_2)} \leq \|A_0\|_{C_p} + \|B\|_{L_a^p(D,\ell_2)}.$$

Proof.   Since $A = A_0 + S(B)$, where S is the unilateral shift, that

is $S(C) := \begin{pmatrix} 0 & c_0^1 & c_1^1 & c_2^1 & \cdots \\ 0 & 0 & c_0^2 & c_1^2 & \ddots \\ 0 & 0 & 0 & c_0^3 & \ddots \\ \vdots & \vdots & \vdots & \vdots & \ddots \end{pmatrix}$ for $C = \begin{pmatrix} c_0^1 & c_1^1 & c_2^1 & \cdots \\ 0 & c_0^2 & c_1^2 & \ddots \\ 0 & 0 & c_0^3 & \ddots \\ \vdots & \vdots & \vdots & \ddots \end{pmatrix}$, we find that

$\|A\|_{L_a^p(D,\ell_2)} \leq \|A_0\|_{C_p} + \|B\|_{L_a^p(D,\ell_2)}$. To get the other inequality, note that $A_n = (n+1) \int_0^1 \int_0^{2\pi} A(re^{it}) r^n e^{-int} \frac{r}{\pi} dt dr$ for $n \geq 0$. This implies the estimate $\|A_n\|_{C_p} \leq (n+1)\|A\|_{L_a^p(D,\ell_2)}$. Hence, since $B(r) = \sum_{n=0}^{\infty} A_{n+1} r^n$, we obtain that

$$\left(\int_0^1 \|B(r)\|_{C_p}^p 2dr \right)^{1/p}$$

$$\leq \left(\int_0^{1/2} \|B(r)\|_{C_p}^p 2r dr \right)^{1/p} + \left(\int_{1/2}^1 \|B(r)\|_{C_p}^p 2r dr \right)^{1/p}$$

$$\leq \frac{1}{4^{1/p}} \left(\sum_{n=0}^{\infty} \frac{(n+1)}{2^n} \right) \|A\|_{L_a^p(D,\ell_2)} + 2 \left(\int_{1/2}^1 (\|A_0\|_{C_p} + \|A(r)\|_{C_p})^p r dr \right)^{1/p}.$$

This gives that $\|B\|_{L_a^p(D,\ell_2)} \leq L\|A\|_{L_a^p(D,\ell_2)}$, and taking $K = 1/(L+1)$, the proof is complete.   □

Another interesting result about Bergman-Schatten classes is as follows [14]:

Theorem 6.7. *Let A be an upper triangular matrix, $n \in \mathbb{N}$, $1 \le p < \infty$. Then $A \in L_a^p(D, \ell_2)$ if and only if the function $r \to (1 - r^2)^n A^{(n)}(r) \in L^p(D, \ell_2)$.*

Proof. We prove that for any upper triangular matrix A and $k \ge 0$, the function $(1 - r^2)^k A(r)$ belongs to $L^p(D, \ell_2)$ if and only if $(1 - r^2)^{k+1} A'(r)$ also does. Then a recurrent argument gives the statement.

Note that $(1 - r^2)^{k+1} A'(r) \in L^p(D, \ell_2)$ if and only if

$$\int_0^1 (1 - r^2)^{pk+p} ||rA'(r)||_{C_p}^p 2dr < \infty.$$

We denote $B(r) = rA'(r) = \sum_{n=0}^\infty nA_n r^n$ and observe that for each $r < 1$ we have that $B(r^2) = A(r) * \Lambda(r)$, where Λ is the Toeplitz matrix associated to the function $\lambda(r) = \frac{r}{(1-r)^2}$.

Since

$$M_1(\lambda, r) := \int_0^{2\pi} |\lambda(re^{it})| \frac{dt}{2\pi} = \frac{r}{1 - r^2}$$

and

$$M_p(B, r^2) := \left(\int_0^{2\pi} ||B(r^2 e^{it})||_{C_p}^p \frac{dt}{2\pi} \right)^{1/p} \le M_1(\lambda, r) M_p(A, r),$$

we obtain that

$$\int_0^1 (1-r^2)^{pk+p} ||rA'(r)||_{C_p}^p 2dr = \int_0^1 \int_0^{2\pi} (1-r^2)^{pk+p} ||re^{it}A'(re^{it})||_{C_p}^p \frac{1}{\pi} dt dr$$

$$= \int_0^1 4r^3 (1 - r^4)^{pk+p} M_p^p(B, r^2) dr \le C \int_0^1 r(1 - r^2)^{pk} M_p^p(A, r) dr$$

$$= C \int_0^1 \int_0^{2\pi} (1-r^2)^{pk} ||A(re^{it})||_{C_p}^p \frac{1}{\pi} dt dr = C \int_0^1 (1-r^2)^{pk} ||A(r)||_{C_p}^p 2r dr.$$

Conversely, we take A such that $(1 - r^2)^{k+1} M_p(A', r) \in L^p((0, 1), dr)$. We may, without loss of generality, assume that $\int_0^1 (1 - r)^{(k+1)p} M_p^p(A', r) dr = 1$ and also that $A_0 = 0$.

Since $M_p(A, r) \le \int_0^r M_p(A', s) ds$ we have that

$$\int_0^1 (1 - r^2)^{kp} ||A(r)||_{C_p}^p 2r dr = \int_0^1 (1 - r^2)^{kp} M_p^p(A, r) 2r dr \le$$

$$\int_0^1 2r(1-r^2)^{kp}\left(\int_0^r M_p(A',s)ds\right)^p dr \le C\int_0^1 (1-r)^{kp}\left(\int_0^r M_p(A',s)ds\right)^p dr.$$

For $p = 1$ we get that

$$\int_0^1 (1-r^2)^k \|A(r)\|_{C_1} 2r dr \le C\int_0^1 (1-r)^k \left(\int_0^r M_1(A',s)ds\right)dr$$

$$= \frac{C}{k+1}\int_0^1 (1-s)^{k+1}M_1(A',s)ds = C'.$$

For $p > 1$ we write, for each $t \in (0,1)$,

$$I_t = \int_0^t (1-r)^{kp}\left(\int_0^r M_p(A',s)ds\right)^p dr.$$

Let $u(r) = -\frac{1}{pk+1}(1-r)^{pk+1}$ and $v(r) = \left(\int_0^r M_p(A',s)ds\right)^p$.
Since $u(t)v(t) < 0$ and $v(0) = 0$, we have that

$$I_t = \int_0^t u'(r)v(r)dr \le -\int_0^t u(r)v'(r)dr.$$

Thus,

$$I_t \le \frac{p}{pk+1}\int_0^t (1-r)^{pk+1}M_p(A',r)\left(\int_0^r M_p(A',s)ds\right)^{p-1} dr$$

$$= \frac{p}{pk+1}\int_0^t (1-r)^{k+1}M_p(A',r)(1-r)^{(p-1)k}\int_0^r M_p(A',s)ds)^{p-1} dr.$$

Then the assumption and Hölder's inequality shows that $I_t \le CI_t^{1/p'}$.
Hence, $I_t \le C$ for all t and the proof is complete. $\qquad\square$

Remark. We observe that for $A \in L_a^1(D,\ell_2)$ and $n \in \mathbb{N}$ we have that

$$\int_0^1 \int_0^{2\pi} A(re^{it})r^n e^{-int}\frac{1}{\pi}rdtdr =$$

$$\int_0^1 \int_0^{2\pi} (1-r^2)A'(re^{it})r^{n-1}e^{-i(n-1)t}\frac{1}{\pi}rdtdr = \frac{A_n}{n+1}.$$

See [14]-Proposition 2.6.

Using the above remark it is easy to prove the following result (see [14]-Proposition 2.7):

Proposition 6.8. *If* $A \in L_a^1(D,\ell_2)$, *then*

$$A(re^{it}) = \int_0^1 \int_0^{2\pi} \frac{A(se^{iu})s}{(1-re^{it}se^{-iu})^2}\frac{1}{\pi}duds =$$

$$2\int_0^1 \int_0^\pi \frac{(1-s^2)A(se^{iu})s}{(1-re^{it}se^{-iu})^3}\frac{1}{\pi}duds \text{ for all } 0 \le r < 1.$$

Lemma 6.9. *Let $A \in L_a^2(D, \ell_2)$, $0 \leq r < 1$ and $B \in C_2$. Then the linear functional $F_{r,B}(A) = \mathrm{tr}\, A(r)B^*$ is continuous on $L_a^2(D, \ell_2)$.*

Proof. If A is an upper triangular matrix of finite order and $A(r) = C(r)*$ A, then we consider the function $f_A(r, \theta)$ on D given in the introduction. It is clear that the function above is an holomorphic C_2-valued function on D, and, consequently, the function $z \to ||\sum_{k=0}^{\infty} A_k r^k e^{ik\theta}||_{C_2}$ is subharmonic.

Thus, for $0 < r' \leq 1 - r$ we have that

$$||A(r)||_{C_2}^2 \leq \frac{1}{\pi r'^2} \int_0^{r'} \int_0^{2\pi} ||\sum_{k=0}^{\infty} A_k |r + se^{i\theta}|^k e^{ik\theta}||_{C_2}^2 s \, d\theta \, ds$$

$$\leq \frac{1}{r'^2} \int_0^{r'} \sum_{k=0}^{\infty} ||A_k||_{C_2}^2 (r + s)^{2k} 2s \, ds \leq \text{(since } r' \leq 1 - r)$$

$$\leq \frac{1}{r'^2} \int_0^1 ||A(s)||_{C_2}^2 2s \, ds = \frac{1}{r'^2} ||A||_{L_a^2(D, \ell_2)}^2.$$

By taking above $r' = 1 - r$ we obtain that

$$|F_{r,B}(A)|^2 = |\mathrm{tr}\, A(r)B^*|^2 \leq ||A(r)||_{C_2}^2 \cdot ||B||_{C_2}^2 \leq \frac{||B||_{C_2}^2}{(1-r)^2} \cdot ||A||_{L_a^2(D, \ell_2)}^2,$$

which implies the continuity of $F_{r,B}$. $\qquad\square$

In view of Lemma 6.9 and Riesz Theorem it follows that there is an unique matrix $K_{r,B} \in L_a^2(D, \ell_2)$ such that

$$F_{r,B}(A) = <A, K_{r,B}>_{L_a^2(D, \ell_2)} = 2 \int_0^1 \mathrm{tr}\, A(s)[K_{r,B}(s)]^* s \, ds$$

for all $B \in C_2$, $0 \leq r < 1$ and $A \in L_a^2(D, \ell_2)$.

Let $i, j \in \mathbb{N}$ and let B be the matrix whose entries $b(k, l)$ are

$$b(k, l) := \delta_{ki} \delta_{lj}.$$

Then the above formula yields that:

$$A(r)(i, j) = 2 \int_0^1 \mathrm{tr}\, A(s) K_{r,i,j}(s)^* s \, ds = 2 \int_0^1 \mathrm{tr}\, [A(s)(K_{r,i,j}^* * P(s))] s \, ds$$

for all $i, j \in \mathbb{N}$, where by $K_{r,i,j}$ we denote the matrix $K_{r,B}$ for the above matrix B, $P(s)$ is the Toeplitz matrix associated to the Poisson kernel, that is

$$P(s) = \begin{pmatrix} 1 & s & s^2 & s^3 & \cdots \\ s & 1 & s & s^2 & \ddots \\ s^2 & s & 1 & s & \ddots \\ \vdots & \ddots & \ddots & \ddots & \ddots \end{pmatrix}.$$

Since $A(r)$ is an analytic matrix, we have that

$$(P(r) * A)(i,j) = \int_0^1 \text{tr}(P(s) * A)[P(s) * K_{r,i,j}](2s)ds$$

for all $j \geq i$ and all $0 \leq r < 1$.

It is easy to see that

$$K_{r,i,j} = \begin{cases} \left((j-i+1)r^{j-i}\delta_{i,m}\delta_{j,l}\right)_{l,m=1}^{\infty} & i \leq j, \\ (0)_{l,m=1}^{\infty} & i > j. \end{cases}$$

Definition 6.10. *Let $r \to A(r)$ be an element of $L^2(D, \ell_2)$. Since $\tilde{L}_a^2(D, \ell_2)$ is a closed subspace in the Hilbert space $L^2(D, \ell_2)$, there is an unique orthogonal projection $\tilde{P}$ on $\tilde{L}_a^2(D, \ell_2)$, called Bergman projection. We denote by P the corresponding operator from $L^2(D, \ell_2)$ onto $L_a^2(D, \ell_2)$.*

Proposition 6.11. *For all functions $r \to A(r)$ from $L^2(D, \ell_2)$ and for all $i, j \in \mathbb{N}$ we have that*

$$\{[P(A(\cdot))](r)\}(i,j) = \begin{cases} 2(j-i+1)r^{j-i}\int_0^1 a_{ij}(s)s^{j-i+1}ds & \text{if } i \leq j, \\ 0 & \text{if } i > j. \end{cases}$$

Proof. We have that

$$[P(A(\cdot))](r)(i,j) = F_{r,i,j}(P(A(\cdot))) = \; < P(A(\cdot)), K_{r,i,j} >$$

$$= (\text{since } P \text{ is a selfadjoint projection}) = \; < A, P(K_{r,i,j}) >$$

$$= (\text{since } K_{r,i,j} \text{ is an analytic matrix}) = \; < A, K_{r,i,j} >_{L_a^2(D,\ell_2)}$$

$$= 2\int_0^1 \text{tr}[A(s)K_{r,i,i}^*(s)]sds = 2(j-i+1)r^{j-i}\int_0^1 a_{ij}(s) \cdot s^{j-i+1}ds,$$

if $j \geq i$ and $< A, B >$ means $\text{tr}AB^*$.

If $j < i$, then it is easy to see that $([P(A(\cdot))])(i,j) = 0$. □

Unfortunately the Bergman projection is unbounded on $L^1(D, \ell_2)$, but instead we can consider a version of it.

Let $\alpha > -1$. Then we put

$$K_{r,i,j,\alpha} = \begin{cases} \left(\frac{\Gamma(j-i+2+\alpha)}{(j-i)!\Gamma(2+\alpha)}r^{j-i}\delta_{i,l}\delta_{j,m}\right)_{l,m=1}^{\infty} & i \leq j, \\ 0 & i > j \end{cases}$$

and we have that, for an analytic matrix $A(s) = P(s) * A$,

$$a_{ij}r^{j-i} = (\alpha+1)2\int_0^1 \sum_{k=0}^\infty < A_k, [K_{r,i,j,\alpha}]_k > s^{2k+1}(1-s^2)^\alpha ds \quad \forall\, i,j,$$

where $A = (a_{ij})_{i,j=1}^\infty$.

Then

$$[P_\alpha A(\cdot)](r) = \begin{cases} \frac{(\alpha+1)\Gamma(j-i+2+\alpha)}{(j-i)!\Gamma(\alpha+2)}r^{j-i}(2\int_0^1 a_{ij}(s)s^{j-i+1}(1-s^2)^\alpha ds) & \text{if } j \geq i \\ 0 & \text{if } j < i. \end{cases}$$

Theorem 6.12. P_1 *is a continuous operator (precisely a continuous projection) from* $L^1(D,\ell_2)$ *onto* $L_a^1(D,\ell_2)$.

Proof. By Theorem 1.2, the topological dual of $L^1(D,\ell_2)$ is $L^\infty(D,\ell_2)$ with respect to the duality pair

$$< A(\cdot), B(\cdot) > := \int_0^1 \text{tr } (A(s)[B(s)]^*)2sds,$$

where $A(\cdot) \in L^\infty(D,\ell_2)$, $B(\cdot) \in L^1(D,\ell_2)$.

Now we are looking for the adjoint P_1^* of P_1:

$$< P_1^*A(\cdot), B(\cdot) > = 2\int_0^1 \sum_{i=1}^\infty \sum_{j=1}^\infty (P_1^*A(\cdot))(r)(i,j)\overline{b_{ij}(r)}rdr$$

$$= \sum_{i=1}^\infty \sum_{j=1}^\infty \int_0^1 (P_1^*A(\cdot))(r)(i,j)\overline{b_{ij}(r)}(2r)dr.$$

On the other hand

$$< P_1^*A(\cdot), B(\cdot) > = < A(\cdot), P_1B(\cdot) > = \int_0^1 \text{tr } A(r)(P_1B)^*(r)(2rdr)$$

$$= \sum_{i=1}^\infty \sum_{j=1}^\infty \int_0^1 A(r)(i,j)\overline{(P_1B)(r)(i,j)}(2rdr)$$

$$= \sum_{i=1}^\infty \sum_{j=i}^\infty \frac{\Gamma(j-i+3)}{(j-i)!\Gamma(2)} \left(\int_0^1 [A(s)](i,j)s^{j-i}(2sds)\right) \times$$

$$\left(\int_0^1 \overline{b_{ij}(s)}s^{j-i}(1-s^2)(2sds)\right).$$

We take $B(s)(i,j) = \chi_{I_k}(s)/(\mu(I_k))$ and $B(s)(l,k) = 0$, $(l,k) \neq (i,j)$, $\forall (i,j) \in \mathbb{N} \times \mathbb{N}$, where $I_k \ni r$ is a sequence of intervals such that $\lim_{k\to\infty} \mu(I_k) = 0$, and $d\mu = 2sds$.

By Lebesgue's differentiation theorem we have that

$$
(P_1^* A(\cdot))(r)(i,j) = \begin{cases} \frac{\Gamma(j-i+3)}{(j-i)!\Gamma(2)} r^{j-i}(1-r^2) \int_0^1 A(s)(i,j)s^{j-i}(2sds) & \text{if } j > i, \\ 0 & \text{if } j \leq i, \end{cases}
$$

a.e. for all $r \in [0,1)$.

We show that $P_1^* : L^\infty(D, \ell_2) \to L^\infty(D, \ell_2)$ is a bounded operator. In order to prove this we have to remark that

$$
\|A(r)\|^2_{L^\infty(D,\ell_2)} = \operatorname*{ess\,sup}_{0 \leq r < 1} \|A(r)\|^2_{B(\ell_2)} =
$$

$$
\operatorname*{ess\,sup}_{0 \leq r < 1} \sup_{\sum_{j=1}^\infty |h_j|^2 \leq 1} \sum_{i=1}^\infty |\sum_{j=1}^\infty a_{ij}(r)h_j|^2.
$$

Consequently, since $L^1[0,1]$ has cotype 2, there is a constant $K > 0$ such that

$$
\|P_1^* A(\cdot)\|^2_{L^\infty(D,\ell_2)} =
$$

$$
\operatorname*{ess\,sup}_{0 \leq r < 1} \sup_{\|h\|_2 \leq 1} \sum_{i=1}^\infty |\sum_{j=i}^\infty \int_0^1 h_j r^{j-i} \frac{\Gamma(j-i+3)}{(j-i)!\Gamma(2)}(1-r^2) a_{ij}(s) s^{j-i}(2sds)|^2 =
$$

$$
\operatorname*{ess\,sup}_{0 \leq r < 1} \sup_{\|h\|_{\ell_2} \leq 1} (1-r^2)^2 \sum_{i=1}^\infty |\int_0^1 (\sum_{j=i}^\infty a_{ij}(s)((rs)^{j-i} \frac{\Gamma(j-i+3)}{(j-i)!}) h_j)(2sds)|^2
$$

$$
< K \operatorname*{ess\,sup}_{0 \leq r < 1} (1-r^2)^2 \sup_{\|h\|_{\ell_2} \leq 1} \times
$$

$$
[\int_0^1 (\sum_{i=1}^\infty |\sum_{j=i}^\infty a_{ij}(s)[(rs)^{j-i}(j-i+2)(j-i+1)]h_j|^2)^{1/2}(2sds)]^2.
$$

Since the Toeplitz matrix $C(rs)$ given by the sequence of functions $((c_{ij})(rs)_{i,j=1}^\infty)$, where

$$
c_{ij}(rs) := c_{j-i}(rs) = \begin{cases} (rs)^{j-i}(j-i+2)(j-i+1) & \text{if } j \geq i \\ 0 & \text{if } j < i, \end{cases}
$$

is a Schur multiplier (we remark that $\sum_{k=0}^{\infty} u^k (k+2)(k+1)e^{ik\theta} = \frac{2}{(1-ue^{i\theta})^3}$), then, by Bennett's Theorem, the multiplier norm of the matrix $C(rs)$ is exactly the $L^1(\mathbb{T})$-norm of $\frac{2}{(1-rse^{i\theta})^3}$, that is it is equal to $2\sum_{n=0}^{\infty} \frac{\Gamma(n+3/2)^2}{(n!)^2\Gamma(3/2)^2}(rs)^{2n}$.

Thus,

$$\sup_{\sum_{j=1}^{\infty}|h_j|^2 \leq 1} \left(\sum_{i=1}^{\infty} \left| \sum_{j=i}^{\infty} a_{ij}(s)(rs)^{j-i}(j-i+2)(j-i+1)h_j \right|^2 \right)^{1/2}$$

$$= ||A(s) * C(rs)||_{B(\ell_2)} \leq ||A(s)||_{B(\ell_2)} \cdot \sum_{n=0}^{\infty} \frac{\Gamma(n+3/2)^2}{(n!)^2\Gamma(3/2)^2}(rs)^{2n}.$$

Consequently,

$$||P_1^* A(\cdot)||^2_{L^{\infty}(D,\ell_2)} \leq$$

$$K \operatorname*{ess\,sup}_{r<1}(1-r^2)^2 \left[\int_0^1 ||A(s)||_{B(\ell_2)} \cdot \sum_{n=0}^{\infty} \frac{\Gamma(n+3/2)^2}{(n!)^2\Gamma(3/2)^2} r^{2n} s^{2n+1}(2ds) \right]^2 \leq$$

$$K \operatorname*{ess\,sup}_{r<1}(1-r^2)^2 ||A(\cdot)||^2_{L^{\infty}(D,\ell_2)} \left(\sum_{n=0}^{\infty} \frac{\Gamma(n+3/2)^2 r^{2n}}{(n!)^2(n+1)\Gamma(3/2)^2} \right)^2 \sim$$

(by Stirling's formula) $\sim \operatorname*{ess\,sup}_{r<1}(1-r^2)^2 \frac{1}{(1-r^2)^2}||A(\cdot)||^2_{L^{\infty}(D,\ell_2)}$

$$\sim ||A(\cdot)||^2_{L^{\infty}(D,\ell_2)},$$

which shows in turn that $P_1^* : L^{\infty}(D,\ell_2) \to L^{\infty}(D,\ell_2)$ is bounded. The proof is complete. $\quad\square$

We have the following duality theorem:

Theorem 6.13. *Let $1 < p < \infty$ and $1/p + 1/q = 1$. Then $(L_a^p(D,\ell_2))^* \sim L_a^q(D,\ell_2)$ with respect to the duality bilinear mapping:*

$$< A(\cdot), B(\cdot) > = \int_0^1 tr[A(r)B^*(r)](2rdr).$$

Proof. We use the boundedness of the projection $P : L^p(D,\ell_2) \to L_a^p(D,\ell_2)$. Since the proof is similar to that of Theorem 7.11 we leave out the details. Cf, also [14] Theorem 3.6. $\quad\square$

6.2   Some inequalities in Bergman-Schatten classes

The following Hardy-Littlewood inequalities are useful in the study of classical Hardy spaces H^p, with $1 \le p \le \infty$. See [23].

Hardy-Littlewood Theorem. *(i) If $0 < p \le 2$, then $f \in H^p$ implies $\sum n^{p-2}|a_n|^p < \infty$, and*

$$\left(\sum_{n=0}^{\infty} (n+1)^{p-2}|a_n|^p \right)^{1/p} \le K_p \|f\|_{H^p},$$

where K_p denotes a constant depending only on p and $f(z) = \sum_{n=0}^{\infty} a_n z^n$, for all $|z| < 1$.

(ii) If $2 \le p \le \infty$, then $\sum n^{p-2}|a_n|^p < \infty$ implies $f \in H^p$, and

$$\|f\|_{H^p} \le K_p \left(\sum_{n=0}^{\infty} (n+1)^{p-2}|a_n|^p \right)^{1/p}.$$

In the book [28] it is given a more complete form of Hardy-Littlewood theorem for classical Bergman spaces. In what follows we give the matrix version of these last inequalities, more precisely we state and prove the matrix version of Theorem 3-p.83 [28]. See also [14]-Proposition 4.1.

We denote by $L_a^{p,\,unc}(D, \ell_2)$ the space

$$\{A|\,\|A\|_{p,unc} := \left(\int_0^1 \int_0^1 \|\sum_{k \ge 0} \epsilon_k(t) A_k r^k\|_{C_p}^p\, r dr dt \right)^{1/p} < \infty\},$$

where ϵ_k means the kth Rademacher function.

Theorem 6.14. *1. If $p = 1$ and $A \in L_a^1(D, \ell_2)$, then $\sum_{n=1}^{\infty} n^{-2}\|A_n\|_{C_1} < \infty$.*

2. If $1 < p \le 2$ and $A \in L_a^{p,unc}(D, \ell_2)$, then we have that $\sum_{n=1}^{\infty} n^{p-3}\|A_n\|_{C_p}^p < \infty$.

3. If $2 \le p < \infty$ and $\sum_{n=1}^{\infty} n^{p-3}\|A_n\|_{C_p}^p < \infty$, then $A \in L_a^{p,unc}(D, \ell_2)$.

4. If $2 \le p < \infty$ and $A \in L_a^p(D, \ell_2)$, then $\sum_{n=1}^{\infty} n^{-1}\|A_n\|_{C_p} < \infty$.

5. If $1 < p \le 2$ and $\sum_{n \ge 0} \frac{1}{n+1}\|A_n\|_{C_p}^p < \infty$, then $A \in L_a^p(D, \ell_2)$.

6. If $1 \le p < \infty$ and $A \in L_a^p(D, \ell_2)$, then $\|A_n\|_{C_p} = o(n^{1/p})$.

Proof. 1. Let $A \in L_a^1(D, \ell_2)$. We use the results of Shields [84] applied to the function $z \to A(rz)$, with a fixed $0 < r < 1$. Then, we find that

$$\sum_{n=0}^{\infty} (n+1)^{-1}\|A_n r^n\|_{C_1} \le C\{M_1(A, r)\}.$$

Multiplying the above inequality by r and integrating over $(0,1)$ we have that

$$\sum_{n=0}^{\infty}(n+1)^{-1}||A_n||_{C_1}\int_0^1 r^{n+1}dr \le ||A||_{L_a^1},$$

or

$$\sum_{n=0}^{\infty}(n+1)^{-2}||A_n||_{C_1} \le C||A||_{L_a^1}.$$

2. Let $A \in A^{p,unc}$ and $1 < p \le 2$. By Theorem 4.5 applied for $A_r(z) = A(rz)$, $0 < r < 1$, with $A_r \in T_p(\ell_R^2) + T_p(\ell_C^2)$, we find that

$$\sum_{n=0}^{\infty}(n+1)^{p-2}||A_n r^n||_{C_p}^p \le C(p)\int_0^1 ||\sum_{k=0}^{\infty}\epsilon_k(t)A_k r^k||_{C_p}^p dt.$$

Proceeding as in the case 1 we get that

$$\sum_{n=0}^{\infty}(n+1)^{p-2}||A_n||_{C_p}^p\int_0^1 r^{np+1}dr \le C(p)||A||_{p,unc}^p,$$

or

$$\sum_{n=0}^{\infty}(n+1)^{p-3}||A_n||_{C_p}^p \le C(p)||A||_{p,unc}^p.$$

3. By Theorem 4.7 applied to $A_r(z) = A(rz)$, for $0 < r < 1$, we have that

$$||A_r||_{p,unc}^p \le C(p)\sum_{n=0}^{\infty}(n+1)^{p-2}||A_n r^n||_{C_p}^p,$$

and, consequently,

$$\int_0^1 ||\sum_{k=0}^{\infty}\epsilon_k(t)A_k r^k||_{C_p}^p dt \le C(p)\sum_{n=0}^{\infty}(n+1)^{p-2}||A_n r^n||_{C_p}^p.$$

Hence, denoting $C(p) = C$,

$$\int_0^1\left[\int_0^1 ||\sum_{k=0}^{\infty}\epsilon_k(t)A_k r^k||_{C_p}^p r dr\right]dt \le C\sum_{n=0}^{\infty}(n+1)^{p-2}||A_n||_{C_p}^p\int_0^1 r^{np+1}dr$$

and finally

$$||A||_{p,unc}^p \le C(p)\sum_{n=0}^{\infty}(n+1)^{p-3}||A_n||_{C_p}^p.$$

4. In the same way as in Corollary 7.6 we get that $L_a^p(D,\ell_2)$ coincides with the interpolation space $[L_a^2(D,\ell_2),\mathcal{B}(D,\ell_2)]_\theta$ for $2/p = 1-\theta$. Moreover,

let $T : \mathcal{B}(D, \ell_2) \to \ell_\infty(\ell_\infty)$; be given by $T(A) = (A_n)_{n \geq 0}$. We note that T maps $L_a^2(D, \ell_2)$ continuously into $\ell_2(\ell_2, w)$, where $w(n) = 1/(n+1)$, for all $n \geq 0$, and (ℓ_2, w) is the space of all sequences $x = (x_n)_{n \geq 0}$ such that $\sum_{n \geq 0} |x_n|^2 w(n) < \infty$, with the natural norm.

Indeed, similarly to Theorem at page 84-[28] we get that

$$||T(A)||^2 = \sum_{n \geq 0} \frac{1}{n+1} ||A_n||_\infty^2 \leq C(2) ||A||_{L_a^2(D, \ell_2)}^2.$$

Hence, $T : L_a^p(D, \ell_2) \to [\ell_2(\ell_2, w), \ell_\infty(\ell_\infty)]_\theta = \ell_p(\ell_p, w)$, where $\theta = 1 - 2/p$, is a bounded operator. (See [93] for the last equation.)

Hence,

$$\left(\sum_{n \geq 0} \frac{1}{n+1} ||A_n||_{C_p}^p \right)^{1/p} \leq C(p) ||A||_{L_a^p(D, \ell_2)}.$$

5. By Theorem 6.13 we find that $[L_a^p(D, \ell_2)]^* = L_a^q(D, \ell_2)$, $1/p + 1/q = 1$. On the other hand, $[\ell_p(\ell_p, w)]^* = \ell_q(\ell_q, w)$, where $1/p + 1/q = 1$. Therefore, the conclusion follows from 4.

6. For each n and $r \in (0, 1)$ we have that

$$A_n r^n = \frac{1}{2\pi} \int_0^{2\pi} A(re^{it}) e^{-int} dt.$$

This implies that, for any $n \in \mathbb{N}$ and $0 < r < 1$, it yields that

$$||A_n||_{C_p} r^n \leq M_1(A, r). \tag{6.1}$$

Since $M_p(A, \cdot)$ is increasing in $(0, 1)$, from (6.1) it follows that

$$(1-r)||A_n||_{C_p}^p r^{np} \leq (1-r) M_p^p(A, r) \leq \int_r^1 M_p^p(A, s) ds$$

for each $r \in (0, 1)$.

Hence, for any n, by taking $r = 1 - \frac{1}{n}$, we see that

$$\frac{1}{n} ||A_n||_{C_p}^p \approx \frac{1}{n} ||A_n||_{C_p}^n \left(1 - \frac{1}{n}\right)^{np} \leq \int_{1-1/n}^1 M_p^p(A, s) ds.$$

This shows that $\frac{||A_n||_{C_p}}{n^{1/p}} \to 0$. $\square$

Remark. In connection with the above results we mention that O. Blasco showed in [14]-Theorem 4.4 that, for $1 \leq p < \infty$, there exist $K_1, K_2 > 0$ such that

$$K_1 \left(\sum_{k=0}^\infty \sup_{2^{k-1} \leq n < 2^k} ||A_n||_{C_p}^p \right)^{1/p}$$

$$\leq ||A||_{L_a^p(D,\ell_2)} \leq K_2 \left(\sum_{k=0}^{\infty} (\sum_{2^{k-1} \leq n < 2^k} ||A_n||_{C_p})^p \right)^{1/p} .$$

We are now ready to present the Hausdorff-Young Theorem for Bergman-Schatten classes extending Theorem 2-[28] at pages 81-82.

Theorem 6.15. *Let* $1 \leq p \leq \infty$ *and let* $q = \frac{p}{p-1}$ *be its conjugate exponent.*
(i) If $1 \leq p \leq 2$, *then* $A \in L_a^p(D,\ell_2)$ *implies that*

$$\left(\sum_{n \geq 0} (n+1)^{1-q} ||A_n||_{T_p}^q \right)^{1/q} \leq ||A||_{L_a^p(D,\ell_2)}.$$

(ii) If $2 \leq p \leq \infty$, *then* $\sum_{n \geq 0} (n+1)^{1-q} ||A_n||_{T_p}^q < \infty$ *implies that* $A \in L_a^p(D,\ell_2)$ *and* $||A||_{L_a^p(D,\ell_2)} \leq \left(\sum_{n=0}^{\infty} (n+1)^{1-q} ||A_n||_{T_p}^q \right)^{1/q}$.

Proof. In case (ii), let μ be the discrete measure on the set $\mathbb{N}$ which assigns the mass $\mu(n)+1$ to the integer $n = 0, 1, 2, \ldots$ Consider the linear operator T that maps the sequence $\{\frac{1}{n+1} A_n\}$ to the formal series $\sum_{n \geq 0} A_n z^n$. We want to show that T is bounded as an operator from $L^q(\mathbb{N}, d\mu; T_p)$ to $L^p(D, d\sigma; T_p)$, with norm $||T|| \leq 1$, where $L^p(D, d\sigma; T_p)$ is the space of all p-Lebesgue Bochner integrable T_p-valued functions defined on the unit disk D with respect to the area measure $d\sigma$ on D.

In the case $p = 2$ this follows from

$$||T(\{A_n/(n+1)\})||_{L^2(D,d\sigma;T_2)}^2 = ||\{A_n/(n+1)\}||_{L^2(\mathbb{N},d\mu;T_2)}^2.$$

For $p = \infty$ it is the clear that $\sup_{|z| \leq 1} ||\sum_n A_n z^n||_{T_\infty} \leq \sum_n ||A_n||_{T_\infty}$.

The result follows by using these estimates and complex interpolation of vector-valued L^p spaces.

Case (i). Let $A(z) \in L^p(D, d\sigma; T_p)$. Define the linear operator $T(A) = \{b_n\}$, where $b_n = \int_D A(z) \overline{z^n} d\sigma$. If $A \in L_a^p(D, \ell_2)$, then $b_n = A_n/(n+1)$. With the measure μ defined as before, we want to show that T is a bounded operator from $L^p(D, d\sigma; T_p)$ to $L^q(\mathbb{N}, d\mu; T_p)$, with norm $||T|| \leq 1$. For $p = 1$ this is the trivial inequality $||b_n||_{T_1} \leq ||A||_{L^1(D,d\sigma;T_1)}$ for all $n \in \mathbb{N}$. For $p = 2$ it follows from the relation

$$\sum_{n=0}^{\infty} (n+1) ||b_n||_{T_2}^2 = \sum_{n=0}^{\infty} (n+1) || \int_D A(z) \overline{z^n} d\sigma ||_{L_a^2(D,\ell_2)}^2$$

$$= \sum_{k=0}^{\infty} \sum_{l=1}^{\infty} \frac{1}{k+1} \int_D \int_D A(z)_k^l \overline{A(\zeta)_k^l} \sum_{n=0}^{\infty} (n+1)(\overline{z}\zeta)^n d\sigma(z) d\sigma(\zeta)$$

$$= \sum_{k=0}^{\infty} \sum_{l=1}^{\infty} \frac{1}{k+1} \int_D \int_D \frac{A(z)_k^l \overline{A(\zeta)_k^l}}{(1-\overline{z}\zeta)^2} d\sigma(z) d\sigma(\zeta) = \; <PA(\cdot), A(\cdot)>,$$

where P denotes the Bergman projection. Since

$$| <PA(\cdot), A(\cdot)>| \leq ||PA(\cdot)||_{L_a^2(D,\ell_2)} ||A(\cdot)||_{L_a^2(D,\ell_2)} \leq ||A(\cdot)||_{L_a^2(D,\ell_2)}^2,$$

this gives the desired inequality for $p = 2$.

By using these estimates and complex interpolation as before we obtain that

$$\left(\sum_{n=0}^{\infty} (n+1)||b_n||_{T_p}^q \right)^{1/q} \leq \left(\int_D ||A(z)||_{T_p}^p d\sigma \right)^{1/p}, \quad A(\cdot) \in L^p(D, d\sigma; T_p).$$

Specializing to $A \in L_a^p(D, \ell_2)$ and recalling that $b_n = A_n/(n+1)$, we arrive at the desired result. $\qquad\square$

6.3   A characterization of the Bergman-Schatten space

We give a characterization of the space $L_a^1(D, \ell_2)$ completely similar to those obtained by Mateljevic and Pavlovic in [61].

First we prove some necessary Lemmas, which are also of independent interest.

Lemma 6.16. *Let* $A = \sum_{k=m}^{n} A_k$, *where* $m \leq n$. *Then*

$$||A||_{C_1} r^n \leq ||A(r)||_{C_1} \leq ||A||_{C_1} r^m,$$

for all $0 < r < 1$.

Proof. We put $B[r] = \sum_{k=m}^{n} \overline{A}_{-k} r^{n-k}$. Then

$$||A(r)||_{C_1} = ||A(r)^*||_{C_1} = || \sum_{k=m}^{n} \overline{A}_{-k} r^k ||_{C_1} = ||B[\frac{1}{r}]||_{C_1} r^n.$$

Since $\frac{1}{r} > 1$ and the function $se^{it} \to ||B[se^{it}]||_{C_1}$, for $0 < s < 1$ and $t \in [0, 2\pi]$, is a subharmonic function we get that

$$||B[\frac{1}{r}]||_{C_1} = \int_0^{2\pi} ||B[\frac{1}{r}](e^{it})||_{C_1} \frac{dt}{2\pi} \geq \int_0^{2\pi} ||B[1](e^{it})||_{C_1} \frac{dt}{2\pi} = ||B[1]||_{C_1}$$

$$= ||A^*||_{C_1} = ||A||_{C_1},$$

so that the left hand side inequality holds.

The other inequality can be proved similarly. The proof is complete. $\square$

Let $\sigma_n(A) = \sum_{k=0}^{n}\left(1 - \frac{k}{n+1}\right)A_k$ be the Cesaro mean of the order n of the upper triangular matrix A, and $||\sigma_n(A)||_1 := \sup_{0<r<1}||\sigma_n(A)(r)||_{C_1}$, for $n = 0, 1, 2, \ldots$

Lemma 6.17. *Then we have*

$$||\sigma_k(A)||_1 r^k \leq ||A(r)||_{C_1} \leq (1-r)^2 \sum_{n=0}^{\infty}||\sigma_n(A)||_1(n+1)r^n.$$

Proof. Since $||\sum_{k=-n}^{n}(1 - \frac{|k|}{n+1})e^{ikt}||_{L^1(\mathbb{T})} \leq 1$ for all $n \in \mathbb{N}$, it follows that $||A(r)||_{C_1} \geq ||\sigma_n(A)(r)||_{C_1}$ for all $0 < r < 1$ and for all $n \in \mathbb{N}$.

Using this inequality and Lemma 6.16 we get the left hand side inequality of Lemma 6.17.

For the right hand side of the inequality of the Lemma we use the elementary formula: $A(r) = (1-r)^2 \sum_{n=0}^{\infty}\sigma_n(A)(n+1)r^n$. The proof is complete. □

Lemma 6.18. *Let A be an upper triangular matrix. Then, for $0 \leq k < n$, we have that*

$$(n - k + 1)||\sigma_k(A)||_1 \leq (n+1)||\sigma_n(A)||_1.$$

Proof. Since by [17] we have that $||\sigma_k(B)||_{C_1} \leq ||B||_{C_1}$ it follows that

$$||\sigma_n(A)||_1 \geq ||\sigma_k\sigma_n(A)||_1 = ||\sigma_k(A) - \frac{r}{n+1}\sigma_k'(A)||_1 \geq ||\sigma_k(A)||_1$$

$$-\frac{1}{n+1}||\sigma_k'(A)||_1.$$

Now it is clear that Bernstein's inequality

$$||\sum_{k=0}^{n}kA_k||_1 \leq n||\sum_{k=0}^{n}A_k||_1$$

holds. Therefore,

$$||\sigma_n(A)||_1 \geq ||\sigma_k(A)||_1 - \frac{k}{n+1}||\sigma_k(A)||_1.$$

The proof is complete. □

We aim to state and prove an analogue of Theorem 2.2 in [61]. In order to do that we recall some results about functions, which were presented in Section 4 in [61].

Let ϕ be a non-negative increasing function defined on $(0,1]$ for which

$$\phi(tr) \le Ct^\alpha \phi(r), \quad 0 < t < 1, \tag{6.2}$$

and

$$\phi(tr) \ge C^{-1}t^\beta \phi(r), \quad 0 < t < 1, \tag{6.3}$$

where C and β are positive real numbers. Note that $\beta \ge \alpha > 0$.

Lemma 6.19. *Let* $\psi(r) = \phi(r)^q r^{-\epsilon}$, *where* $q < \infty$, $q\alpha - \epsilon > -1$, *and let* α *satisfy the condition (6.2). Then*

$$C^{-1}x^{-1}\psi(1/x) \le \int_0^1 \psi(1-r)r^{x-1}dr \le Cx^{-1}\psi(1/x), \quad x \ge 1.$$

Proof. We have that

$$I(x) := \int_0^1 \psi(1-r)r^{x-1}dr = x^{-1}\int_0^x \psi(t/x)(1-t/x)^{x-1}dt.$$

Since ϕ satisfies the conditions (6.2) and (6.3) it yields that

$$\phi(t/x) \le C(t^\alpha + t^\beta)\phi(1/x), \quad 0 < t < x \quad (\beta \ge \alpha),$$

and, consequently,

$$I(x) \le Cx^{-1}\psi(1/x)\int_0^x (t^\alpha + t^\beta)^q t^{-\epsilon}(1-t/x)^{x-1}dt$$

$$\le Cx^{-1}\psi(1/x).$$

This proves the right-hand side inequality. The left-hand side inequality is easy and does not depend on the conditions (6.2) and (6.3). The proof is complete. □

Lemma 6.20. *Let* $\psi(r) = \phi(r)r^{-\epsilon}$, $\epsilon \le \alpha$ *and let* α *satisfy (6.2). Then*

$$C^{-1}\psi(1/x) \le \sup_r \psi(1-r)r^{x-1} \le C\psi(1/x), \quad x \ge 1.$$

Proof. The proof is similar to that of Lemma 6.19. □

Besides these lemmas we shall use the well-known estimate

$$\sum_{n=0}^{\infty} 2^{n\beta}r^{2^n} \le Cr(1-r)^{-\beta}, \quad \beta > 0. \tag{6.4}$$

The proof of Theorem 6.22, which follows, is based on L^q-behaviour of the functions

$$F_1(r) = (1-r)^{-1/q}\phi(1-r)\sup\{\lambda_n r^{2^n} : n \ge 0\}$$

and

$$F_2(r) = (1 - r)^{-1/q}\phi(1 - r)\sum_{n=0}^{\infty}\lambda_n r^{2^n},$$

where (λ_n) is a sequence of non-negative real numbers.

Proposition 6.21. *Let $F = F_1$ or $F = F_2$. Then*

$$C^{-1}\|F\|_{L^q} \leq \|(\phi(2^{-n})\lambda_n)\|_{\ell_q} \leq \|F\|_{\ell_q}.$$

Proof. We consider only the case $q < \infty$. In the case $q = \infty$ the proof is similar and is based on Lemma 6.20.

Let $q < \infty$. Then

$$F(r)^q \geq F_1(r)^q \geq (1 - r)^{-1}\phi(1 - r)^q\lambda_k^q r^{2^k q} \quad \text{for all } k$$

and, by (6.4),

$$F(r)^q \geq C^{-1}\phi(1 - r)^q\sum_{n=0}^{\infty}2^n r^{2^n}\lambda_k r^{2^k q}.$$

Hence,

$$F(r)^q \geq C^{-1}\phi(1 - r)^q\sum_{n=0}^{\infty}2^n\lambda_n^q r^{2^n(1+q)}. \tag{6.5}$$

On the other hand, from Lemma 3.4 and hypothesis (6.3) it follows that

$$\int_0^1 \phi(1 - r)^q r^{2^n(q+1)}dr \geq C^{-1}2^{-n}\phi^q(2^{-n}).$$

Combining this with (6.5), we obtain the right-hand side inequality in Proposition 3.6.

To prove the left-hand side inequality, let

$$\eta_n = 2^{n\delta}r^{2^{n-1}}, \quad \theta_n = 2^{-n\delta}\lambda_n r^{2^{n-1}},$$

where $\delta = \alpha/2$ and α satisfies (6.2). Then

$$\left(\sum_{n=0}^{\infty}\lambda_n r^{2^n}\right)^q = \left(\sum_{n=0}^{\infty}\eta_n\theta_n\right)^q \leq \left(\sum_{n=0}^{\infty}\eta_n\right)^q\sum_{n=0}^{\infty}\theta_n^q$$

$$\leq C(1 - r)^{-q\delta}\sum_{n=0}^{\infty}2^{-nq\delta}\lambda_n^q r^{2^{n-1}q},$$

where we have used (6.4). Hence,

$$F(r)^q \leq F_2(r)^q \leq C\psi(1 - r)\sum_{n=0}^{\infty}2^{-nq\delta}\lambda_n^q r^{2^{n-1}q},$$

where $\psi(r) = \phi(r)^q r^{-q\delta-1}$. Now the desired result follows from Lemma 6.19. □

Our main result in this section reads:

Theorem 6.22. *Let A be an upper triangular matrix. Then $A \in L^1_a(D, \ell_2)$ if and only if*

$$\sum_{n=1}^{\infty} \frac{||\sigma_n(A)||_1}{(n+1)^2} < \infty.$$

Proof.   First we remark that $A \in L^1_a(D, \ell_2)$ if and only if

$$\int_0^1 ||A(r)||_{C_1} dr < \infty.$$

Let $A \in L^1_a(D, \ell_2)$. Then, by the left hand side inequality of Lemma 6.17,

$$||A(r)||_{C_1} = (1-r) \sum_{n=0}^{\infty} ||A(r)||_{C_1} r^n \geq$$

$$(1-r) \sum_{n=0}^{\infty} ||\sigma_n(A)||_1 r^{2n}.$$

Consequently, an integration with respect to r yields

$$\infty > \int_0^1 ||A(r)||_{C_1} dr \geq \int_0^1 (1-r) \sum_{n=0}^{\infty} ||\sigma_n(A)||_1 r^{2n} dr =$$

$$\sum_{n=0}^{\infty} \frac{1}{(2n+1)(2n+2)} ||\sigma_n(A)||_1 \geq \frac{1}{8} \sum_{n=0}^{\infty} \frac{1}{(n+1)^2} ||\sigma_n(A)||_1.$$

Conversely, suppose that

$$\sum_{n=0}^{\infty} \frac{1}{(n+1)^2} ||\sigma_n(A)||_1 < \infty.$$

Let

$$x_n = \sum_{k=0}^{n} (k+1)(n-k+1) ||\sigma_k(A)||_1.$$

Then, summing by parts as in (4.4) in [61], we get that

$$\sum_{n=0}^{\infty} ||\sigma_n(A)||_1 (n+1) r^n = (1-r)^2 \sum_{n=0}^{\infty} x_n r^n. \tag{6.6}$$

On the other hand, using Lemma 6.18, we see that

$$x_n \leq C(n+1)^3 ||\sigma_n(A)||_1$$

and, therefore,

$$\sum_{n=0}^{\infty} \frac{1}{(n+1)^5} x_n \leq C \sum_{n=0}^{\infty} \frac{1}{(n+1)^2} ||\sigma_n(A)||_1 < \infty.$$

By Lemma 4.8 in [61], with $q = 1$, $\phi(r) = r$, $r \in (0, 1]$, we get that

$$\int_0^1 (1-r)^4 \sum_{n=0}^{\infty} x_n r^n dr < \infty.$$

Using (6.6) we arrive at

$$\int_0^1 (1-r)^2 \sum_{n=0}^{\infty} ||\sigma_n(A)||_1 (n+1) r^n dr < \infty$$

and, by the right hand side inequality in Lemma 6.17, finally we obtain that

$$\int_0^1 ||A(r)||_{C_1} dr < \infty$$

i.e. $A \in L_a^1(D, \ell_2)$. The proof is complete. $\qquad\square$

6.4 Usual multipliers in Bergman-Schatten spaces

In this section we characterize the multipliers with respect to usual product of matrices for the Bergman-Schatten spaces of index 2, $L_a^2(\ell_2)$.

More specific, let A be an infinite upper triangular matrix. A is called an *usual multiplier for* $L_a^2(\ell_2)$ and one writes $A \in M(L_a^2)$, if for all $B \in L_a^2(\ell_2)$, it follows that $AB \in L_a^2(\ell_2)$, or equivalently (by theorem of closed graph) there is the smallest constant $M(A) < \infty$ such that

$$||A \cdot B||_{L_a^2(\ell_2)} \leq M(A)||B||_{L_a^2(\ell_2)}. \tag{6.7}$$

Now let $b \in \mathbb{N}$. Let us denote by $\ell_2^n(w) = \{x = (x_1, x_2, \ldots, x_n)$ with the norm $||x||_{2,w} \overset{def}{=} \left(\sum_{j=1}^{n} \frac{|x_j|^2}{n-j+1}\right)^{1/2}\}$.

Then, A being as above we denote by $||A||_n = ||A||_{B(\ell_2^n(w), \ell_2^n(w))}$, $n \in \mathbb{N}$. We have the following result:

Theorem 6.23. *A is an usual multiplier for* $L_a^2(\ell_2)$ *if and only if*

$$\sup_n ||A||_n < \infty \text{ and } M(A) = \sup_n ||A||_n.$$

Proof.   First we denote by $P(r)$ the Toeplitz matrix

$$
\begin{pmatrix}
1 & r & r^2 & r^3 & \cdots \\
r & 1 & r & r^2 & \cdots \\
r^2 & r & 1 & r & \cdots \\
\vdots & \ddots & \ddots & \ddots & \ddots
\end{pmatrix}, \, 0 \le r < 1
$$

and we denote by $A(r) = A * P(r)$, where, as usual in this book $*$ means the Schur product of matrices.

Let us remark that $A(r) \cdot B(r) = (A \cdot B)(r)$ for all B upper triangular matrices and for all $r \in [0, 1)$.

Then, for an usual multiplier A for $L_a^2(\ell_2)$, we have

$$
\int_0^1 \| \sum_{k=0}^{\infty} (AB)_k r^k \|_{C_2}^2 (2r dr) \le M(A)^2 \int_0^1 \| \sum_{k=0}^{\infty} B_k r^k \|_{C_2}^2 2r dr, \qquad (6.8)
$$

for all upper triangular matrices B. Take now $B = B^{(n)}$, the nth column of B. Then, substituting in (6.8) we have, for $A = \left(a_k^j \right)_{k \ge 0, j \ge 1}$ and $B^{(n)} =$

$$
\begin{pmatrix}
0 \, 0 \, \ldots \, 0 \, b_{n-1}^1 \, 0 \, \ldots \\
0 \, 0 \, \ldots \, 0 \, b_{n-1}^2 \, 0 \, \ldots \\
\vdots \, \vdots \, \vdots \quad \vdots \, \vdots \qquad \vdots \, \vdots \\
0 \, 0 \, \ldots \, 0 \, b_0^n \quad 0 \, \ldots \\
0 \, 0 \, \ldots \, 0 \, 0 \qquad 0 \, \ldots
\end{pmatrix},
$$

$$
AB^{(n)}
$$

$$
= \begin{pmatrix}
a_0^1 & a_1^1 & a_2^1 & a_3^1 & \ldots & a_{n-1}^1 & \cdots \\
0 & a_0^2 & a_1^2 & a_2^2 & \ldots & a_{n-2}^2 & \cdots \\
0 & 0 & a_0^3 & a_1^3 & \ldots & a_{n-3}^3 & \cdots \\
\vdots & \vdots & \vdots & \vdots & \vdots & \vdots & \vdots \\
0 & 0 & & \ldots & \ldots & a_0^n & \cdots \\
0 & 0 & & \ldots & \ldots & 0 & \cdots
\end{pmatrix} \cdot \begin{pmatrix}
0 \, 0 \, \ldots \, 0 \, b_{n-1}^1 \, 0 \, \ldots \\
0 \, 0 \, \ldots \, 0 \, b_{n-2}^2 \, 0 \, \ldots \\
0 \, 0 \, \ldots \, 0 \, b_{n-3}^3 \, 0 \, \ldots \\
\vdots \, \vdots \, \vdots \quad \vdots \, \vdots \qquad \vdots \, \vdots \\
0 \, 0 \, \ldots \, 0 \, b_0^n \quad 0 \, \ldots \\
0 \, 0 \, \ldots \, 0 \, 0 \qquad 0 \, \ldots
\end{pmatrix}
$$

$$
= \begin{pmatrix}
0 \, 0 \, \ldots \, 0 \, a_0^1 b_{n-1}^1 + a_1^1 b_{n-2}^2 + \cdots + a_{n-1}^1 b_0^n \, 0 \, \ldots \\
0 \, 0 \, \ldots \, 0 \, a_0^2 b_{n-2}^2 + a_1^2 b_{n-3}^3 + \cdots + a_{n-2}^2 b_0^n \, 0 \, \ldots \\
\vdots \, \vdots \, \vdots \quad \vdots \, \vdots \qquad\qquad\qquad\qquad\qquad\quad \vdots \, \vdots \\
0 \, 0 \, \ldots \, 0 \, a_0^n b_0^n \qquad\qquad\qquad\qquad\qquad 0 \, \ldots \\
0 \, 0 \, \ldots \, 0 \, 0 \qquad\qquad\qquad\qquad\qquad\qquad 0 \, \ldots
\end{pmatrix}.
$$

Hence

$$
\| AB^{(n)} \|_{L_a^2(\ell_2)}^2 =
$$

$$\int_0^1 \left(|\sum_{j=0}^{n-1} a_j^1 b_{n-1-j}^{j+1}|^2 r^{2n-2} + |\sum_{j=0}^{n-2} a_j^2 b_{n-2-j}^{j+2}|^2 r^{2n-4} + \cdots + |a_0^n b_0^n|^2 \right) (2r\,dr)$$

$$= |\sum_{j=0}^{n-1} a_j^1 b_{n-1-j}^{j+1}|^2 \cdot \frac{1}{n} + |\sum_{j=0}^{n-2} a_j^2 b_{n-2-j}^{j+2}|^2 \cdot \frac{1}{n-1} + \cdots + |a_0^n b_0^n|^2,$$

and

$$||B^{(n)}||^2_{L_a^2(\ell_2)} = |b_{n-1}^1|^2 \cdot \frac{1}{n} + |b_{n-2}^2|^2 \cdot \frac{1}{n-1} + \cdots + |b_0^n|^2.$$

Hence

$$\left(|\sum_{j=0}^{n-1} a_j^1 b_{n-1-j}^{j+1}|^2 \cdot \frac{1}{n} + |\sum_{j=0}^{n-2} a_j^2 b_{n-2-j}^{j+2}|^2 \cdot \frac{1}{n-1} + \cdots + |a_0^n b_0^n|^2 \right)^{1/2} \le$$

$$M(A) \left(\sum_{j=0}^{n-1} |b_j^{n-j}|^2 \cdot \frac{1}{n-j} \right)^{1/2} = || \left(b_{n-j}^j \right)_{j=1}^n ||_{\ell_2^n(w)}$$

or

$$||A||_n \le M(A)$$

for all n. Consequently

$$\sup_n ||A||_n \le M(A).$$

Conversely, if $\sup_n ||A||_n = M < \infty$, if k is arbitrary but fixed and $B = \sum_{k=0}^{n-1} B^{(k)} \in L_a^2(\ell_2)$, we have, denoting by $\left(AB^{(k)}\right)^{(k)}$ the sequence placed on the kth column beginning from the top,

$$||AB||^2_{L_a^2(\ell_2)} = \sum_{k=1}^n ||AB^{(k)}||^2_{L_a^2(\ell_2)} = \sum_{k=1}^n || \left(AB^{(k)}\right)^{(k)} ||^2_{\ell_2^k(w)}$$

$$\le M \sum_{k=1}^n || \left(B^{(k)}\right) ||^2_{\ell_2^k(w)} \le M \sum_{k=1}^n ||B^{(k)}||^2_{L_a^2(\ell_2)} = M ||B||^2_{L_a^2(\ell_2)}$$

for all k.

Hence

$$AB \in L_a^2(\ell_2) \ \forall \ B \in L_a^2(\ell_2) \text{ and } M(A) = \sup_n ||A||_n.$$

$\square$

Now we can characterize the matrices A such that $\sup_n \|A\|_n < \infty$.

Theorem 6.24. *If A is an upper triangular matrix, then it follows that $\sup_n \|A\|_n < \infty$ if and only if $A \in B(\ell_2)$.*

Proof. Let $A = \left(a_k^l\right)_{l \geq 1, k \geq 0}$ such that $\sup_n \|A\|_n = 1 < \infty$.
Then, for a column matrix B such that

$$(B) = \begin{pmatrix} b_{n-1}^1 \\ b_{n-2}^2 \\ \vdots \\ b_0^n \end{pmatrix} \in \ell_2^n(w)$$

with

$$\|(B)\|_{2,w}^2 = \frac{|b_{n-1}^1|^2}{n} + \frac{|b_{n-2}^2|^2}{n-1} + \cdots + |b_0^n|^2 = 1$$

we have

$$\|(AB)\|_{2,w}^2 = \frac{|\sum_{k=0}^{n-1} a_k^1 b_{n-k-1}^{k+1}|^2}{n} + \frac{|\sum_{k=0}^{n-2} a_k^2 b_{n-k-1}^{k+2}|^2}{n-1} + \cdots + |a_0^n b_0^n|^2 =$$

$$\frac{|\sum_{k=0}^{n-1}(a_k^1 \sqrt{n-k}) \frac{b_{n-k-1}^{k+1}}{\sqrt{n-k}}|^2}{n} + \frac{|\sum_{k=0}^{n-2}(a_k^2 \sqrt{n-k-1}) \frac{b_{n-k-2}^{k+2}}{\sqrt{n-k-1}}|^2}{n-1} + \cdots + |a_0^n b_0^n|^2.$$

If we denote by $y_k = \frac{b_{n-k-1}^{k+1}}{\sqrt{n-k}}$, where $k = 0, 1, \ldots, n-1$ we have
$\|(B)\|_{2,w}^2 = |y_0|^2 + \cdots + |y_{n-1}|^2 = \|(y_k)\|_{\ell_2}^2$.
Hence

$$\|(AB)\|_{2,w}^2 = |\sum_{k=0}^{n-1} a_k^1 \sqrt{\frac{n-k}{n}} y_k|^2 + |\sum_{k=0}^{n-2} a_k^2 \sqrt{\frac{n-k-1}{n-1}} y_{k+1}|^2 + \cdots + |a_0^n y_{n-1}|^2$$

$$= \|A_n' y\|_{\ell_2}^2,$$

where

$$A_n' = \begin{pmatrix} a_0^1 & a_1^1\sqrt{\frac{n-1}{n}} & a_2^1\sqrt{\frac{n-2}{n}} & \cdots & \frac{a_{n-1}^1}{\sqrt{n}} & 0 \cdots \\ 0 & a_0^2 & a_2^1\sqrt{\frac{n-2}{n-1}} & \cdots & \frac{a_{n-2}^2}{\sqrt{n-1}} & 0 \cdots \\ 0 & 0 & a_0^3 & \cdots & \frac{a_{n-3}^3}{\sqrt{n-2}} & 0 \cdots \\ \vdots & \vdots & \vdots & \vdots & \vdots & \vdots \cdots \\ 0 & 0 & 0 & \cdots & a_0^n & 0 \cdots \end{pmatrix}.$$

Hence $||A_n||_{B(\ell_2)} \leq ||A||_n$, which in turn implies that $\sup_n ||A'_n||_{B(\ell_2)} \leq \sup_n ||A||_n = 1$, in other words the sequence $(A'_n)_{n=1}^{\infty}$ belongs to the unit ball $B(B(\ell_2))$ of $B(\ell_2)$.

It is easy to see that $A'_n(k,l) \to A(k,l)$ for all $k,l \in \mathbb{N}$. Since the unit ball $B(B(\ell_2))$ is $\sigma(B(\ell_2), C_1)$-compact it follows that $A \in B(B(\ell_2))$, that is $||A||_{B(\ell_2)} \leq 1 = \sup_n ||A||_n$.

The converse implication is easy, it is enough to verify that such a matrix is an usual multiplier on $L_a^2(\ell_2)$ and apply the previous theorem. □

Each matrix $\Phi \in L_a^2(\ell_2)$ is said to *generate* a subspace $[\Phi]$, the closure of the set of multiples of Φ by matrices of finite band type. When $\Phi \in H^\infty(\ell_2)$, it is important to observe that $[\Phi]$ does not necessarily coincide with
$$\Phi L_a^2(\ell_2) = \{\Phi F : F \in L_a^2(\ell_2)\},$$
since the latter set need not be closed. The crucial requirement is that the operator of multiplication by Φ be *bounded below*: in other words, that there exist a constant $c > 0$ such that $||\Phi F||_{L_a^2(\ell_2)} \geq c||F||_{L_a^2(\ell_2)}$ for all $F \in L_a^2(\ell_2)$.

Theorem 6.25. *Let Φ be an invertible matrix from $H^\infty(\ell_2)$. Then the set $\Phi L_a^2(\ell_2)$ is closed in $L_a^2(\ell_2)$ if and only if the operator of multiplication M_Φ is bounded below on $L_a^2(\ell_2)$.*

Proof. Suppose first that M_Φ is bounded below. Then if $A^n \in L_a^2(\ell_2)$ and $\{\Phi A^n\}$ converges to some $G \in L_a^2(\ell_2)$, it follows that $A^n \to A$ for some $A \in L_a^2(\ell_2)$. Hence $\Phi A^n \to \Phi A$ because $\Phi \in H^{\infty(\ell_2)}$ is an usual multiplier on $L_a^2(\ell_2)$. Hence $G = \Phi A \in \Phi L_a^2(\ell_2)$, which is consequently a closed subspace.

Consequently, we define $T(G) = \Phi^{-1}G$ with the domain $\mathcal{D}(T) = \{G \in [\Phi] : \Phi^{-1}G \in L_a^2(\ell_2)\}$.

It is easy to remark that T is a closed operator. Indeed if $G^n \to G \in L_a^2(\ell_2)$, where $G^n \in \mathcal{D}(T)$ and $T(g^n) \to F$, that is $\Phi^{-1}G^n \to F \in L_a^2(\ell_2)$, then it follows that $G^n \to M_\Phi(F) = \Phi F$ in $L_a^2(\ell_2)$. Consequently $G = \Phi F$ and $\Phi^{-1}G = F \in L_a^2(\ell_2)$, that is $G \in \mathcal{D}(T)$ and T is a closed operator.

Observe first that $\Phi L_a^2(\ell_2)$ is closed if and only if $\Phi L_a^2(\ell_2) = [\Phi]$, because the matrices of finite band type are dense in $L_a^2(\ell_2)$.

Since by hypothesis $\Phi L_a^2(\ell_2)$ is a closed subspace and $\mathcal{D}(T) = \{G = \Phi A, A \in L_a^2(\ell_2), \Phi^{-1}G \in L_a^2(\ell_2)\} = \Phi L_a^2(\ell_2)$, by closed graph theorem it follows that T is a bounded operator.

Hence $||A||_{L_a^2(\ell_2)} \leq C||\Phi A||_{L_a^2(\ell_2)}$ for some constant $C > 0$ and for all $A \in L_a^2(\ell_2)$. Hence M_Φ is a bounded below operator. □

Now we give some interesting examples of matrices Φ such that M_Φ be bounded below operator on $L_a^2(\ell_2)$.

Let $\alpha \in \mathbb{C}$ with $|\alpha| < 1$, $\alpha \neq 0$. Let $0 \leq r_0 < |\alpha|$ and $\Phi = \Phi_\alpha *$ $C(r_0)$, where Φ_α is the Toeplitz matrix associated to the Moebius transform $\phi_\alpha(z) = \frac{\alpha - z}{1 - \overline{\alpha}z}$, $|z| < 1$.

Then M_Φ is bounded below linear and bounded operator on $L_a^2(\ell_2)$.

Indeed we have

$$
\Phi_\alpha = \begin{pmatrix}
\alpha & |\alpha|^2 - 1 & (|\alpha|^2 - 1)\overline{\alpha} & (|\alpha|^2 - 1)\overline{\alpha}^2 & \cdots \\
0 & \alpha & |\alpha|^2 - 1 & (|\alpha|^2 - 1)\overline{\alpha} & \cdots \\
0 & 0 & \alpha & |\alpha|^2 - 1 & \ddots \\
0 & 0 & 0 & \alpha & \ddots \\
\vdots & \vdots & \vdots & \vdots & \ddots
\end{pmatrix}.
$$

Taking $0 \leq r < 1$ and $B \in L_a^2(\ell_2)$ we have

$$
[M_\Phi(B)](r) = \begin{pmatrix}
\alpha & (|\alpha|^2 - 1)r_0 r & (|\alpha|^2 - 1)\overline{\alpha}r_0^2 r^2 & (|\alpha|^2 - 1)\overline{\alpha}^2 r_0^3 r^3 & \cdots \\
0 & \alpha & (|\alpha|^2 - 1)r_0 r & (|\alpha|^2 - 1)\overline{\alpha}r_0^2 r^2 & \cdots \\
0 & 0 & \alpha & (|\alpha|^2 - 1)r_0 r & \ddots \\
0 & 0 & 0 & \alpha & \ddots \\
\vdots & \vdots & \vdots & \vdots & \ddots
\end{pmatrix}.
$$

$$
\begin{pmatrix}
b_0^1 & b_1^1 r & b_2^1 r^2 & b_3^1 r^3 & \cdots \\
0 & b_0^2 & b_1^2 r & b_2^2 r^2 & \cdots \\
0 & 0 & b_0^3 & b_1^3 r & \ddots \\
0 & 0 & 0 & b_0^4 & \ddots \\
\vdots & \vdots & \vdots & \vdots & \ddots
\end{pmatrix} =
$$

$$
\alpha B(r) - (1 - |\alpha|^2)r_0\tau_1[(B - B^1)(r)] - (1 - |\alpha|^2)\overline{\alpha}r_0^2\tau_2[(B - B^1 - B^2)(\overline{\alpha}r_0 r)] -
$$

$$
(1 - |\alpha|^2)\overline{\alpha}^2 r_0^3\tau_3[(B - B^1 - B^2 - B^3)(r)] - \cdots
$$

Here B^k is the kth row of B and τ_k is the upper translation with k rows of the corresponding matrix.

Then it follows that

$$
\|M_\Phi(B)\|_{L_a^2(\ell_2)} = \left(\int_0^1 \|M_\Phi(B)(r)\|_{C_2}^2 (2r\,dr) \right)^{1/2} \leq \left[|\alpha| + (1 - |\alpha|^2)r_0 + \right.
$$

$$(1 - |\alpha|^2)r_0^2|\overline{\alpha}| + (1 - |\alpha|^2)r_0^3|\overline{\alpha}|^2 + \dots\Big]\,||B||_{L_a^2(\ell_2)} =$$

$$\left[|\alpha| + (1 - |\alpha|^2)r_0\frac{1}{1 - r_0|\alpha|}\right]||B||_{L_a^2(\ell_2)} \le [|\alpha| + (1 + |\alpha|)r_0]\,||B||_{L_a^2(\ell_2)} =$$

$$C(\alpha)||B||_{L_a^2(\ell_2)},$$

hence *the multiplication operator* M_Φ *is bounded on* $L_a^2(\ell_2)$ (even if r_0 is not less than $|\alpha|$).

Now we will show that M_Φ, *if* $0 \le r_0 < |\alpha|$ *is bounded below on* $L_a^2(\ell_2)$. Indeed

$$||\Phi_\alpha^{-1} * C(r_0)||_{B(\ell_2)} \le K(\alpha).$$

More specific

$$\Phi_\alpha^{-1} = \begin{pmatrix} \frac{1}{\alpha} & \frac{1-|\alpha|^2}{\alpha^2} & \frac{1-|\alpha|^2}{\alpha^3} & \cdots \\ 0 & \frac{1}{\alpha} & \frac{1-|\alpha|^2}{\alpha^2} & \cdots \\ 0 & 0 & \frac{1}{\alpha} & \cdots \\ \vdots & \vdots & \vdots & \ddots \end{pmatrix}.$$

Let $0 \le r_0 < |\alpha|$.
Then

$$\Phi_\alpha^{-1} * C(r_0) = \begin{pmatrix} \frac{1}{\alpha} & \frac{1-|\alpha|^2}{\alpha}(\frac{r_0}{\alpha}) & \frac{1-|\alpha|^2}{\alpha}(\frac{r_0}{\alpha})^2 & \cdots \\ 0 & \frac{1}{\alpha} & \frac{1-|\alpha|^2}{\alpha}(\frac{r_0}{\alpha}) & \cdots \\ 0 & 0 & \frac{1}{\alpha} & \cdots \\ \vdots & \vdots & \vdots & \ddots \end{pmatrix}.$$

Hence

$$||\Phi_\alpha^{-1} * C(r_0)||_{B(\ell_2)} \le \frac{1}{|\alpha|} + \frac{1-|\alpha|^2}{|\alpha|}\left\|\begin{pmatrix} 0 & \frac{r_0}{\alpha} & (\frac{r_0}{\alpha})^2 & \cdots \\ 0 & 0 & \frac{r_0}{\alpha} & \cdots \\ \vdots & \vdots & \vdots & \ddots \end{pmatrix}\right\|_{B(\ell_2)} =$$

$$\frac{1}{|\alpha|} + \frac{1-|\alpha|^2}{|\alpha|}\frac{r_0}{|\alpha|}\sup_{t\in\mathbb{R}}\frac{1}{|1 - \frac{r_0}{|\alpha|}e^{it}|} \le \frac{1}{|\alpha|} + \frac{1-|\alpha|^2}{|\alpha|}\frac{r_0}{|\alpha|}\frac{1}{1 - \frac{r_0}{|\alpha|}} =$$

$$\frac{1-|\alpha|r_0}{|\alpha| - r_0} = \frac{1}{\phi_{|\alpha|}(r_0)} = K(\alpha, r_0).$$

Since

$$\left[\Phi_\alpha^{-1} * C(r_0)\right]\left[\Phi_\alpha * C(r_0)\right] = \left(\Phi_\alpha^{-1}\Phi_\alpha\right) * C(r_0) = I * C(r_0) = I,$$

it follows that

$$\Phi_\alpha^{-1} * C(r_0) = [\Phi_\alpha * C(r_0)]^{-1}.$$

But

$$B * C(r) = \left\{ [\Phi_\alpha * C(r_0)]^{-1} * C(r) \right\} \{\Phi_\alpha * C(r_0) * C(r)\} \{B * C(r)\} =$$

$$\left\{ \left[\Phi_\alpha^{-1} * C(r_0) \right] \Phi B \right\} * C(r).$$

Hence

$$\|B\|_{L_a^2(\ell_2)}^2 = \int_0^1 \|B * C(r)\|_{C_2}^2 2r dr =$$

$$\int_0^1 \| \left\{ \left[\Phi_\alpha^{-1} * C(r_0) \right] \Phi B \right\} * C(r)\|_{C_2}^2 2r dr =$$

$$\| \left[\Phi_\alpha^{-1} * C(r_0) \right] \Phi B \|_{L_a^2(\ell_2)}^2 \leq (\text{by Theorems 4.2 and 4.3}) \leq$$

$$\|\Phi_\alpha^{-1} * C(r_0)\|_{B(\ell_2)}^2 \|\Phi B\|_{L_a^2(\ell_2)}^2 \leq K(\alpha, r_0)^2 \|\Phi B\|_{L_a^2(\ell_2)}^2.$$

Notes

In Section 6.1 we introduce a version of matrix valued Bergman spaces studied previously independently by O. Blasco [14]. We call these spaces Bergman-Schatten spaces. They are appropriate spaces in order to develop a theory similar to classical Harmonic Analysis. For instance see Theorem 6.7 which is a perfect analogue of Theorem 4.2.9 [94].

Moreover, a matrix version of Bergman Projection is introduced and a duality theorem between Bergman-Schatten spaces (Theorem 6.13) is given.

In Section 6.2 we prove some inequalities similar with those from the monograph [28]. In particular, Theorem 6.15 is the Hausdorff-Young Theorem for Bergman Schatten classes.

In Section 6.3 we derive a characterization of the Bergman-Schatten space (see Theorem 6.22) which is completely similar with that obtained by Mateljevic and Pavlovic in [61].

In Section 6.4 we characterize the usual multipliers on the Bergman-Schatten space of index 2.

Chapter 7

A matrix version of Bloch spaces

7.1 Elementary properties of Bloch matrices

The Bloch functions and the Bloch space have a long history behind them. They were introduced by the French mathematician André Bloch in the beginning of the last century. Many mathematicians paid attention to these spaces e.g. the following: L. Ahlfors, J. M. Anderson, J. Clunie, Ch. Pommerenke, P. L. Duren, B. W. Romberg and A. L. Shields. Correspondingly, there are a lot of interesting results in this area (see for example [27], [3], and the following recent monographs [94] and [28]).

Our aim is to introduce the concept of *Bloch matrix,* which extends the notion of Bloch function and to prove some results generalizing those of the earlier cited paper [3].

The basic idea behind our considerations is to consider an infinite matrix A as the analogue of the formal Fourier series associated to a 2π-periodic distribution, the diagonals A_k, $k \in \mathbb{Z}$, being the analogues of the Fourier coefficients associated to the above distribution. In this manner we get a one-to-one correspondence between infinite Toeplitz matrices and formal Fourier series associated to periodic distributions. Hence, an infinite matrix appears in a natural way as a more general concept than those of a periodic distribution on the torus.

Definition 7.1. *The matriceal Bloch space $\mathcal{B}(D, \ell_2)$ is the space of all analytic matrices A with $A(r) \in B(\ell_2)$, $0 \leq r < 1$, such that*

$$||A||_{\mathcal{B}(D,\ell_2)} := \sup_{0 \leq r < 1} (1 - r^2)||A'(r)||_{B(\ell_2)} + ||A_0||_{B(\ell_2)} < \infty,$$

where $B(\ell_2)$ is the usual operator norm of the matrix A on the sequence space ℓ_2, and $A'(r) = \sum_{k=0}^{\infty} A_k k r^{k-1}$.

A matrix $A \in \mathcal{B}(D, \ell_2)$ is called a *Bloch matrix*.

It is clear that the Toeplitz matrices, which belong to the Bloch space of analytic matrices $\mathcal{B}(D, \ell_2)$ coincide with Bloch functions. Hence, $\mathcal{B}(D, \ell_2)$ appears as an extension of the classical Bloch space of functions $\mathcal{B}$.

An important class of Bloch matrices consists of the space $L_a^\infty(D, \ell_2)$ as the following proposition shows:

Proposition 7.2. *The Banach space $L_a^\infty(D, \ell_2)$ is a subspace of $\mathcal{B}(D, \ell_2)$,*

$$\|A\|_{\mathcal{B}(D,\ell_2)} \le 6\|A\|_{L_a^\infty(D,\ell_2)}$$

and $L_a^\infty(D, \ell_2)) \subsetneq \mathcal{B}(D, \ell_2)$.

More precisely, the analytic Toeplitz matrix A given by the sequence $\{\frac{1}{k+1}\}_{k=0}^\infty$ is not in $L_a^\infty(D, \ell_2)$, but $A \in \mathcal{B}(D, \ell_2)$.

Proof. We note that $(1 - r^2)A'(r) = C_1(r) * A^1(r)$, where

$$C_1(r)(k, j) = \begin{cases} (1 - r^2)(j - k)r^{(j-k-1)/2} & \text{if } j \ge k+1, \\ 0 & \text{if } j < k+1, \end{cases}$$

and $A^1(r) = A_0 + \sum_{k=1}^\infty A_k r^{(k-1)/2}$.

But $\|C_1(r)\|_{M(\ell_2)} \le 2$.

Thus,

$$\|(1 - r^2)A'(r)\|_{B(\ell_2)} = \|C_1(r) * A^1(r)\|_{B(\ell_2)} \le 2\|A^1(r)\|_{B(\ell_2)} \quad \forall 0 \le r < 1,$$

implying that

$$\|A\|_{\mathcal{B}(D,\ell_2)} \le 2 \sup_{0 \le r < 1} \|A^1(r)\|_{B(\ell_2)} + \|A_0\|_{B(\ell_2)} \le 3 \sup_{0 \le r < 1} \|A^1(r)\|_{B(\ell_2)}$$

$$\le 9\|A\|_{L_a^\infty(D,\ell_2)}.$$

Moreover, if $A \in L_a^\infty(D, \ell_2)$, then clearly $A \in B(\ell_2)$.

But, taking $h_1 = h_2 = \cdots = h_n = \frac{1}{\sqrt{n}}$ and $h_{n+1} = h_{n+2} = \cdots = 0$, it follows that $\sum_{i=1}^\infty |h_i|^2 = 1$ and

$$\|A\|_{B(\ell_2)}^2 \ge \frac{1}{n}(1 + 1 + \frac{1}{2} + \cdots + \frac{1}{n-1})^2 + \frac{1}{n}(1 + 1 + \frac{1}{2} + \cdots + \frac{1}{n-2})^2$$

$$+ \cdots + \frac{1}{n}(1)^2 \ge C \ln(n-1) \to \infty.$$

On the other hand $A \in \mathcal{B}(D, \ell_2)$.

Indeed, it is easy to see that

$$\sup_{0 \le r < 1} (1 - r^2)\|A'(r)\|_{B(\ell_2)} = \sup_{0 \le r < 1} r^2 = 1.$$

The proof is complete. $\qquad\qquad\qquad\qquad\qquad\qquad\qquad\qquad\qquad\qquad\qquad$ $\square$

The matrix version of Bloch space can be considered as the limit case of $L_a^p(D, \ell_2)$ as $p \to \infty$.

First we note that if $A \in L^\infty(D, \ell_2)$, then $r \to A(r)$ is a w^*-measurable function and, consequently, each function $a_{ij}(r)$ is a Lebesgue measurable function on $[0, 1)$ for all i and j and we may introduce $PA(\cdot)$ as in Proposition 6.11.

Theorem 7.3. $P : L^\infty(D, \ell_2) \to \mathcal{B}(D, \ell_2)$ and $P|_{\widetilde{L^\infty}(D,\ell_2)} : \widetilde{L^\infty}(D, \ell_2) \to \mathcal{B}(D, \ell_2)$ *are bounded surjection operators.*

Proof. Clearly it is enough to prove only the first assertion. Let $A(\cdot) \in L^\infty(D, \ell_2)$ and $B = PA(\cdot)$. We show that $B \in \mathcal{B}(D, \ell_2)$.

It yields that

$$\|B'(r)\|_{B(\ell_2)}^2 \leq$$

$$\sup_{\|h\|_2 \leq 1} \left[\int_0^1 \left(\sum_{i=1}^\infty \left| \sum_{j=i+1}^\infty a_{ij}(s) r^{j-i-1} s^{j-i} (j-i+1)(j-i) h_j \right|^2 \right) (2sds) \right]$$

$$\leq \int_0^1 \|A(s) * C(r, s)\|_{B(\ell_2)}^2 (2sds),$$

where

$$C(r, s)(i, j) = \begin{cases} (j-i+1)(j-i)(rs)^{j-i-1}s & \text{if } j > i, \\ 0 & \text{if } j \leq i. \end{cases}$$

Thus,

$$\|B'(r)\|_{B(\ell_2)} \leq C\|A(\cdot)\|_{L^\infty(D,\ell_2)} \cdot \left(\int_0^1 \left(\int_{-\pi}^\pi \frac{s}{|1 - rse^{i\theta}|^3} \frac{d\theta}{\pi} \right)^2 sds \right)^{1/2}$$

$$\sim \text{(by Lemma 1.3)} \sim C\|A(\cdot)\|_{L^\infty(D,\ell_2)} \cdot \frac{1}{1 - r^2}.$$

Consequently, $\|B\|_{\mathcal{B}(D,\ell_2)} \leq C\|A(\cdot)\|_{L^\infty(D,\ell_2)}$, that is, $P : L^\infty(D, \ell_2) \to \mathcal{B}(D, \ell_2)$ is a bounded operator.

In order to show that P is onto we take $B \in \mathcal{B}(D, \ell_2)$ and put $B^1 = B - B_0 - B_1$. Easy calculations show us that $P[(B^1)'(r)(1 - r^2)] = B^1 * T$, where $T = (t_{j-i})_{i,j \in \mathbb{Z}}$, with $t_k = \frac{4|k|(|k|+1)}{(2|k|-1)(2|k|+1)}$.

Clearly $T^1 = (t_{j-i}^1)_{i,j \in \mathbb{Z}}$, where $t_k^1 = \frac{(2|k|-1)(2|k|+1)}{4|k|(|k|+1)}$, is a Schur multiplier.

Hence, it follows that $B^2(r) := T^1 * (B^1(r))'(1-r^2) \in L^\infty(D,\ell_2)$ and $P[B^2(\cdot)] = B^1$. We only need to prove that $B_0 + B_1 r \in L^\infty(D,\ell_2)$.

Since $P[B_0 + B_1 r] = B_0 + B_1$, it suffices to show that $r \to B_1 r \in L^\infty(D,\ell_2)$, (since $B_0 \in B(\ell_2)$ by the hypothesis $B \in \mathcal{B}(D,\ell_2)$).

Then, since

$$\|B_1\|_{\mathcal{B}(D,\ell_2)} = \sup_{0\le r<1}(1-r^2)\|B_1\|_{B(\ell_2)} = \|B_1\|_{B(\ell_2)},$$

it follows that $B_1 \in B(\ell_2)$.

Thus $r \to B_1 r \in L^\infty(D,\ell_2)$. The proof is complete. $\qquad\square$

Remark 7.4. *Note that $\mathcal{B}(D,\ell_2)$ endowed with the norm $\|\cdot\|_{\mathcal{B}(D,\ell_2)}$ is a Banach space and, by the open mapping theorem, it follows that $\mathcal{B}(D,\ell_2)$ is isomorphic to the quotient space $L^\infty(D,\ell_2)/Ker\,P$, endowed with quotient norm.*

Theorem 7.5. *The projection P_1 is a bounded operator from $L^\infty(D,\ell_2)$ (respectively from $L^\infty(\widetilde{D},\ell_2)$) onto $\mathcal{B}(D,\ell_2)$.*

Proof.   The proof is an easy adaptation of the proof of Theorem 6.12 and thus we leave out the details. $\qquad\square$

By using Theorem 7.5 and Theorem 6.12 we easily get the following corollary:

Corollary 7.6. $L_a^p(D,\ell_2) = [L_a^1(D,\ell_2), \mathcal{B}(D,\ell_2)]_\theta$ *with equivalent norms, for $1 < p < \infty$ and $1 - \theta = 1/p$.*

Indeed we use the known result about the interchangeability of the interpolation functor and a bounded projection. (See [93].)

Now we give some properties of these Bloch matrices, which extend the corresponding properties for Bloch functions.

It is known that in [3] a characterization of Taylor coefficients of Bloch functions in terms of a quadratic form is given. We want to extend this result to the case with infinite matrices.

We recall the definition of the space $\mathcal{I}$ from the paper [3]:

$$\mathcal{I} = \{g : D \to \mathbb{C} \mid \frac{1}{2\pi}\int_0^1\int_0^{2\pi}|g'(z)|d\theta\,dr + |g(0)| < \infty\},$$

equipped with the norm $\|g\|_{\mathcal{I}} := |g(0)| + \frac{1}{2\pi}\int_0^1\int_0^{2\pi}|g'(z)|d\theta\,dr$. Then the following lemma holds:

Lemma 7.7. *Let* $A \in \mathcal{B}(D, \ell_2)$; $A = \sum_{n=0}^{\infty} A_n z^n$ *and* $g(z) = \sum_{n=0}^{\infty} b_n z^n \in$
$\mathcal{I}$. *Then* $h(z) = \sum_{n=0}^{\infty} A_n b_n z^n : \overline{D} \to \mathcal{B}(\ell_2)$ *is a continuous function in*
$|z| \leq 1$ *and, moreover,*

$$\|h(z)\|_{\mathcal{B}(\ell_2)} \leq 2\|A\|_{\mathcal{B}(D,\ell_2)}\|g\|_{\mathcal{I}} \tag{7.1}$$

for all $\|z\| \leq 1$.

In particular, it follows that there exists

$$\langle A, g \rangle = \lim_{\rho \to 1^-} \sum_{n=0}^{\infty} A_n b_n \rho^n = \lim_{\rho \to 1^-} \frac{1}{2\pi} \int_0^{2\pi} A(\rho e^{-i\theta}) g(\rho e^{i\theta}) d\theta,$$

for all $f \in \mathcal{B}(D, \ell_2)$, $g \in \mathcal{I}$.

Proof. Let $|\zeta| < 1$. We have that

$$A'(z) = f_A'(z) = \sum_{n=1}^{\infty} n A_n z^{n-1}$$

and

$$\frac{d}{dz}[z(g(z) - b_0)] = \sum_{n=1}^{\infty} (n+1) b_n z^n.$$

Then, we get easily that, for $z = r e^{i\theta}$ and $\zeta \in D$,

$$\frac{1}{\pi} \int_0^1 \int_0^{2\pi} (1 - r^2) A'(\zeta \bar{z}) \frac{d}{dz}[z(g(z) - b_0)] e^{-i\theta} d\theta \, dr = \sum_{n=1}^{\infty} A_n b_n \zeta^{n-1}.$$

By applying Hölder's inequality and integrating term by term we find that

$$\left\| \sum_{n=1}^{\infty} A_n b_n \zeta^n \right\|_{\mathcal{B}(\ell_2)}$$

$$\leq \sup_{|z|<1} (1 - |z|^2) \|A'(\zeta \bar{z})\|_{\mathcal{B}(\ell_2)} \frac{1}{\pi} \int_0^1 \int_0^{2\pi} (|g(z) - b_0| + r|g'(z)|) d\theta \, dr.$$

Furthermore, we have that

$$\int_0^1 \int_0^{2\pi} |g(re^{i\theta}) - b_0| d\theta \, dr \leq \int_0^1 \int_0^{2\pi} \int_0^r |g'(te^{i\theta})| dt d\theta \, dr =$$

$$\int_0^{2\pi} \int_0^1 \left(\int_t^1 dr \right) |g'(te^{i\theta})| dt d\theta = \int_0^{2\pi} \int_0^1 (1 - t)|g'(te^{i\theta})| dt d\theta.$$

Since $z \to \|A'(z)\|_{B(\ell_2)}$ is a subharmonic function, we get that

$$\| \sum_{n=0}^{\infty} A_n b_n \zeta^n \|_{B(\ell_2)} \leq \|A_0 b_0\|_{B(\ell_2)}$$

$$+ \sup_{z \in D}(1 - |z|^2)\|A'(z)\|_{B(\ell_2)} \frac{1}{\pi} \int_0^1 \int_0^{2\pi} |g'(te^{i\theta})|d\theta dt.$$

Hence,

$$\|h(\zeta)\|_{B(\ell_2)} \leq 2\|A\|_{\mathcal{B}(D,\ell_2)}\|g\|_{\mathcal{I}},$$

for $|\zeta| < 1$.

In order to show the continuity of h in $|z| \leq 1$, we take $\zeta_1, \zeta_2 \in D$ and note that

$$\|h(\zeta_1) - h(\zeta_2)\|_{B(\ell_2)} = \| \sum_{n=0}^{\infty} A_n(b_n\zeta_1^n - b_n\zeta_2^n)\|_{B(\ell_2)} \leq$$

$$2\|A\|_{\mathcal{B}(D,\ell_2)}\|g(\zeta_1) - g(\zeta_2)\|_{\mathcal{I}}.$$

But it is known that the last norm converges to 0 as $|\zeta_1 - \zeta_2| \to 0$. (See Theorem 2.2 [3].)

Hence, h can be extended by continuity to $\overline{D}$ and we get (7.1). The proof is complete. $\qquad\square$

Theorem 7.8. *Let $A = \sum_k A_k$ be a Bloch matrix.*
Then the following inequality holds:

$$\| \sum_{\mu=0}^{\infty} \sum_{\nu=0}^{\infty} \frac{A_{\mu+\nu+1}}{\mu+\nu+1} w_\mu w_\nu \|_{B(\ell_2)} \leq K \sum_{\nu=0}^{\infty} \frac{|w_\nu|^2}{2\nu+1}, \qquad (7.2)$$

where w_ν, $\nu = 0, 1, 2, \ldots$ are complex numbers and $K \leq 2\|A\|_{\mathcal{B}(D,\ell_2)}$.
Conversely, (7.2) implies that $A \in \mathcal{B}(D, \ell_2)$ and $\|A\|_{\mathcal{B}(D,\ell_2)} \leq 2K$.

Proof. It is clear that the double series converges if the right hand series converges.

Therefore we have that

$$\sum_{\mu=0}^{\infty} \sum_{\nu=0}^{\infty} \frac{A_{\mu+\nu+1}}{\mu+\nu+1} w_\mu w_\nu = \sum_{n=0}^{\infty} \frac{A_{n+1}}{n+1} \left(\sum_{\nu=0}^{n} w_\nu w_{n-\nu} \right) = \langle A, g \rangle, \qquad (7.3)$$

where

$$g(z) = \sum_{n=0}^{\infty} \frac{1}{n+1} \left(\sum_{\nu=0}^{n} w_\nu w_{n-\nu} \right) z^{n+1}, \quad z \in D$$

and by $\langle A, g \rangle$ we mean that

$$\langle A, g \rangle = \lim_{\rho \to 1^-} \sum_{n=0}^{\infty} A_n b_n \rho^n = \lim_{\rho \to 1^-} \frac{1}{2\pi} \int_0^{2\pi} A(\rho e^{-i\theta}) g(\rho e^{i\theta}) d\theta,$$

for $A(\rho e^{-i\theta}) = \sum_{n=0}^{\infty} A_n \rho^n e^{-in\theta}$ and $g(\rho e^{i\theta}) = \sum_{n=0}^{\infty} b_n \rho^n e^{in\theta}$, where the series are $\|\cdot\|_{B(\ell_2)}$-convergent.

Moreover,

$$g'(z) = \sum_{n=0}^{\infty} (\sum_{\nu=0}^{n} w_\nu w_{n-\nu}) z^n = (\sum_{n=0}^{\infty} w_n z^n)^2.$$

Hence, by the Parseval formula, we get that

$$\|g\|_{\mathcal{I}} = \frac{1}{2\pi} \int_0^1 \int_0^{2\pi} |g'(z)| d\theta \, dr = \int_0^1 \frac{1}{2\pi} \int_0^{2\pi} |\sum_{n=0}^{\infty} w_n z^n|^2 d\theta \, dr =$$

$$\int_0^1 \sum_{n=0}^{\infty} |w_n|^2 r^{2n} dr = \sum_{n=0}^{\infty} \frac{|w_n|^2}{2n+1}.$$

Therefore, by Lemma 6.7, we have that

$$\| <A, g> \|_{B(\ell_2)} \leq 2\|A\|_{\mathcal{B}(D,\ell_2)} \sum_{n=0}^{\infty} \frac{|w_n|^2}{2n+1},$$

that is (7.2) holds for $A \in \mathcal{B}(D, \ell_2)$.

b) Conversely, if (7.2) holds we take $\zeta \in D$ and find w_n such that $g(z) = \sum_{n=1}^{\infty} n\zeta^{n-1} z^n = \frac{z}{(1-\zeta z)^2} = \sum_{n=0}^{\infty} \frac{1}{n+1} (\sum_{\nu=0}^{n} w_\nu w_{n-\nu}) z^n$. (See Theorem 3.5-[3].)

Using the computations done in [3] at page 17 it follows that

$$\sum_{n=0}^{\infty} \frac{|w_n|^2}{2n+1} = \|g\|_{\mathcal{I}} \leq \frac{2}{1-|\zeta|^2}.$$

By (7.2) and (7.3) we get that

$$\|A'(\zeta)\|_{B(\ell_2)} = \| <A(z), \frac{z}{(1-z\zeta)^2} > \|_{B(\ell_2)} \leq K \sum_{n=0}^{\infty} \frac{|w_n|^2}{2n+1} \leq \frac{2K}{1-|\zeta|^2}.$$

Therefore

$$\|A\|_{\mathcal{B}(D,\ell_2)} \leq 2K.$$

The proof is complete. $\qquad\qquad\qquad\qquad\qquad\qquad\qquad\qquad\qquad\qquad \Box$

We can give an elementary class of Bloch matrices. (See also Proposition 1.5 in [14].)

Theorem 7.9. *Let* $A = \sum_{k=0}^{\infty} A_{2^k}$. *Then* $A \in \mathcal{B}(D, \ell_2)$ *if and only if* $\sup_k \|A_k\|_{B(\ell_2)} < \infty$.

Proof. By Theorem 7.8 it follows that there is a constant $C > 0$ such that $C\|A\|_{\mathcal{B}(D,\ell_2)} \geq \sup_k \|A_k\|_{B(\ell_2)}$, for all infinite matrices A.

We consider a *lacunary* matrix A as in the statement of the theorem. Then

$$\frac{\|zf'_A(z)\|_{B(\ell_2)}}{1 - |z|} = \left(\sum_{n=0}^{\infty} |z|^n \right) \left\| \sum_{k=0}^{\infty} A_{2^k} 2^k z^{2^k} \right\|_{B(\ell_2)} \leq \sup_k \|A_{2^k}\|_{B(\ell_2)} \cdot$$

$$\sum_{n=1}^{\infty} \left(\sum_{2^k \leq n} 2^k \right) |z|^n \leq 2\sup_k \|A_{2^k}\|_{B(\ell_2)} \sum_{n=1}^{\infty} n|z|^n = \frac{2|z|}{(1 - |z|)^2} \sup_k \|A_{2^k}\|_{B(\ell_2)}.$$

Consequently,

$$(1 - r^2)\|A'(r)\|_{B(\ell_2)} \leq 4\sup_k \|A_{2^k}\|_{B(\ell_2)},$$

which, obviously, implies that

$$\|A\|_{\mathcal{B}(D,\ell_2)} \leq 4\sup_k \|A_k\|_{B(\ell_2)}. \qquad \Box$$

It was remarked in [3] that the classical Bloch space of functions $\mathcal{B}$ is a Banach algebra with respect to convolution, or, equivalently, to Hadamard (Schur) composition of functions, that is, for $f = \sum_{k=0}^{\infty} a_k e^{ik\theta} \in \mathcal{B}$ and $g = \sum_{k=0}^{\infty} b_k e^{ik\theta} \in \mathcal{B}$, $f * g = \sum_{k=0}^{\infty} a_k b_k e^{ik\theta} \in \mathcal{B}$. (See [3].)

Now we extend this remark in the framework of matrices with respect to the Schur product. This result was kindly communicated to us by V. Lle.

Theorem 7.10. *The space* $\mathcal{B}(D, \ell_2)$ *is a commutative Banach algebra with respect to Schur product of matrices.*

Proof. Let

$$A = \begin{pmatrix} a_1^0 & a_1^1 & \cdots & \cdots \\ 0 & a_2^0 & a_2^1 & \cdots \\ 0 & 0 & a_3^0 & \cdots \\ \vdots & \vdots & \vdots & \ddots \end{pmatrix}, \quad f_j(re^{it}) = \sum_{k=0}^{\infty} a_j^k r^k e^{ikt},$$

where $0 \leq r < 1$, $t \in [0, 2\pi)$ and $||A||'_{\mathcal{B}(D,\ell_2)} = \sup_{r<1}(1 - r^2)||A'(r)||_{B(\ell_2)}$.

Then, denoting by f'_j the partial derivative of f_j with respect to r, $||A||'_{\mathcal{B}(D,\ell_2)}$ is given by:

$$||A||'_{\mathcal{B}(D,\ell_2)} = \sup_{r<1}(1 - r^2)\{\sup_{||h||_2 \leq 1}(\sum_{j=1}^{\infty}|\int_0^{2\pi} f'_j(re^{it})e^{ijt}h(e^{-it})\frac{dt}{2\pi}|^2)^{1/2}\}.$$

(See [18].)

Hence,

$$(||A * B||'_{\mathcal{B}(D,\ell_2)})^2 =$$

$$\sup_{r<1}\{\sup_{||h||_2 \leq 1}(1 - r^2)^2 \sum_{j=1}^{\infty}|\int_0^{2\pi} (f_j * g_j)'(re^{it})e^{ijt}h(e^{-it})\frac{dt}{2\pi}|^2\},$$

where $(f_j)_j$ corresponds to A as above and $(g_j)_j$ corresponds to B.

Then, we have that

$$r(f_j * g_j)'(re^{it}) = 2\int_0^{\sqrt{r}}\int_0^{2\pi} f'_j(se^{i(\theta+t)})g'_j(se^{-i\theta})s\frac{d\theta}{2\pi}ds,$$

for all j.

By the Cauchy-Schwarz inequality it follows that

$$\sum_{j=1}^{\infty}\left|\int_0^{2\pi} (f_j * g_j)'(re^{it})e^{ijt}h(e^{-it})\frac{dt}{2\pi}\right|^2 =$$

$$4\sum_{j=1}^{\infty}\frac{1}{r^2}\left|\int_0^{\sqrt{r}}\int_0^{2\pi} g'_j(\frac{s}{e^{i\theta}})\frac{s}{e^{ij\theta}}(\int_0^{2\pi} f'_j(se^{i(\theta+t)})e^{ij(t+\theta)}h(\frac{1}{e^{it}})\frac{dt}{2\pi}\frac{d\theta}{2\pi}ds\right|^2$$

$$\leq \sum_{j=1}^{\infty}4r^{-2}\left(\int_0^{\sqrt{r}}\left(\int_0^{2\pi}|g'_j(se^{-i\theta})|^2\frac{d\theta}{2\pi}\right)sds\right) \times$$

$$\left(\int_0^{\sqrt{r}}\int_0^{2\pi} s\left|\int_0^{2\pi} f'_j(se^{i(\theta+t)})e^{ij(t+\theta)}h(e^{-it})\frac{dt}{2\pi}\right|^2\frac{d\theta}{2\pi}ds\right) := I.$$

Moreover,

$$\sup_{j\geq 1}\sup_{s<1}(1 - s^2)^2\left(\int_0^{2\pi}|g'_j(se^{-it})|^2\frac{d\theta}{2\pi}\right) \leq (||B||'_{\mathcal{B}(D,\ell_2)})^2$$

and for $||h||_2 = 1$ we also have that

$$(1 - s^2)^2 \sum_{j=1}^{\infty} \left| \int_0^{2\pi} f_j'(se^{it})e^{ijt}h(e^{-it})\frac{dt}{2\pi} \right|^2 \leq (||A||'_{\mathcal{B}(D,\ell_2)})^2.$$

Consequently,

$$I \leq 4r^{-2} \int_0^{\sqrt{r}} \frac{s(||B||'_{\mathcal{B}(D,\ell_2)})^2}{(1 - s^2)^2}ds \int_0^{\sqrt{r}} \frac{s(||A||'_{\mathcal{B}(D,\ell_2)})^2}{(1 - s^2)^2}ds =$$

$$(1 - r)^{-2}(||A||'_{\mathcal{B}(D,\ell_2)})^2(||B||'_{\mathcal{B}(D,\ell_2)})^2,$$

that is

$$||A * B||'_{\mathcal{B}(D,\ell_2)} \leq C||A||'_{\mathcal{B}(D,\ell_2)}||B||'_{\mathcal{B}(D,\ell_2)}.$$

The proof is complete. □

Theorem 7.11. *The Banach space $L_a^1(D, \ell_2)^*$ (dual of $L_a^1(D, \ell_2)$) may be identified with $\mathcal{B}(D, \ell_2)$. Namely, let $A \in L_a^1(D, \ell_2)$ and $B \in \mathcal{B}(D, \ell_2)$. Then we have that*

$$| < A, B > | = |\int_0^1 tr\,[A(r)B^*(r)](2rdr)| \leq C||A||_{L_a^1(D,\ell_2)} \cdot ||B||_{\mathcal{B}(D,\ell_2)},$$

where $C > 0$ is a constant.

Proof. Since C_1, the Schatten trace class of operators, is a separable Banach space with $C_1^* = B(\ell_2)$, by $< A, B > = tr(AB^*)$, we have, in view Theorem 1.2, that $L^1(D, \ell_2)^* = L^{\infty}(D, \ell_2)$, using the duality map $< A(r), B(r) > = \int_0^1 tr\,[A(r)B^*(r)](2rdr)$.

Then we have that $\tilde{L}_a^1(D, \ell_2)^* = L^{\infty}(D, \ell_2)/(\tilde{L}_a^1(D, \ell_2))^{\perp}$.

Using the fact that $L_a^1(D, \ell_2)$ is canonically isomorphic to $\tilde{L}_a^1(D, \ell_2)$, we have to show that

$$Ker\,P = Ker\,\tilde{P} = (L_a^1(D, \ell_2))^{\perp} \quad \text{in } L^{\infty}(D, \ell_2).$$

But $Ker\,\tilde{P} \subset (\tilde{L}_a^1(D, \ell_2))^{\perp}$, since for $A(r) \in L^{\infty}(D, \ell_2)$ such that $\tilde{P}A(\cdot) = 0$, we have, at least for finitely order matrices $A(\cdot), B(\cdot)$, that

$$< \tilde{P}A(\cdot), B(\cdot) > = < A(\cdot), \tilde{P}B(\cdot) >,$$

and if $B \in \tilde{L}_a^1(D, \ell_2))$, then

$$< A(\cdot), B(\cdot) > = < A(\cdot) - \tilde{P}A(\cdot), B(\cdot) > = < A(\cdot), B(\cdot) - \tilde{P}B(\cdot) > = 0,$$

and, consequently, $A(\cdot) \in (\tilde{L}_a^1(D, \ell_2))^{\perp}$.

Conversely, let $A(\cdot) \in (L_a^1(D, \ell_2))^\perp$, that is $< A(\cdot), B(\cdot) > = 0 \,\forall B \in L_a^1(D, \ell_2))$. Taking $B(r)(i, j) = r^{j-i}$ for $j > i$, with fixed j, i and $B(r)(i, j) = 0$ otherwise, we get that $\int_0^1 a_{ij}(r)(2r dr) = 0$ for all $j > i$. Thus $(\tilde{P}A)(r)(i, j) = 0$ for all i, j, that is $A(\cdot) \in Ker \, \tilde{P}$.

For $B(r) \in L^\infty(D, \ell_2)$ and $A \in L_a^1(D, \ell_2)$, we easily get that

$$| \int_0^1 tr \, [A(r)B^*(r)](2r dr)| \le \int_0^1 |tr \, [A(r)B^*(r)]|(2r dr) \le$$

$$\int_0^1 ||A(r)||_{C_1} \cdot ||B(r)||_{B(\ell_2)} 2r dr \le ||A||_{L_a^1(D,\ell_2)} \cdot ||B(\cdot)||_{L^\infty(D,\ell_2)},$$

so using Remark 7.4 we get the required inequality, since for $A \in L_a^1(D, \ell_2)$, $B \in \mathcal{B}(D, \ell_2)$ we have obviously that

$$| < A, B > | = | < A(r), B(r) > | \le ||A||_{L_a^1(D,\ell_2)} ||B(\cdot)||_{L^\infty(D,\ell_2)}$$

$\forall \, B(\cdot) \in L^\infty(D, \ell_2)$ such that $PB(\cdot) = B$. The proof is complete. $\square$

Lemma 7.12. *Let A be a matrix of finite band-type, that is $A = \sum_{k=1}^n A_k$, such that $A_k \in C_1$ for $k = 1, 2, \ldots$ and let $B \in \mathcal{B}(D, \ell_2)$. Then*

$$< A, B > = \sum_{k=0}^\infty \frac{1}{k+1} tr \, (A_k * \overline{B}_k).$$

Proof. We recall that $< A, B > = \int_0^1 tr \, [A(r)B^*(r)]2r dr$. We denote the entries of the diagonal matrix A_k, where $k \in \mathbb{Z}$, by $(a_k^l)_{l=1}^\infty$. Then it is easy to see that

$$tr \, A(r)B^*(r) = \sum_{l=1}^n \left(\sum_{k=0}^\infty a_k^l \overline{b_k^l} r^{2k} \right)$$

and, consequently,

$$< A, B > = \int_0^1 \sum_{l=1}^n \left(\sum_{k=0}^\infty a_k^l \overline{b_k^l} r^{2k} \right) 2r dr = \int_0^1 \sum_{k=0}^\infty 2r^{2k+1} \left(\sum_{l=1}^n a_k^l \overline{b_k^l} \right) dr$$

$$= \sum_{k=0}^\infty \frac{1}{k+1} \left(\sum_{l=1}^n a_k^l \overline{b_k^l} \right) = \sum_{k=0}^\infty \frac{1}{k+1} tr \, (A_k * \overline{B}_k).$$

The proof is complete. $\square$

Next we present a result, which shows us the intrinsic connection between the matricial Bloch space and the Bergman metric. (See Thm. 5.1.6 [94] for the classical Toeplitz (function) case.)

Theorem 7.13. *If $A \in \mathcal{B}(D, \ell_2)$, then*

$$\sup_{0 \leq r < 1} (1 - r^2) \|A'(r)\|_{B(\ell_2)} = \sup_{z,w \in D \, z \neq w} \frac{\|f_A(z) - f_A(w)\|_{B(\ell_2)}}{\beta(z, w)}.$$

Proof. The proof is simply a translation of the proof of Theorem 5.1.6 [94], replacing f by f_A. We omit the details. □

In particular, we conclude that, in fact, the Bloch matrices are those upper triangular matrices such that $B * C(z)$ are Lipschitz $B(\ell_2)$-valued functions with respect to Bergman metric on D.

We can use this result for proving that $\mathcal{B}(D, \ell_2)$ is a Banach space. We note that by Theorem 7.11 this is clear, but the next proof is much simpler.

Theorem 7.13 has as an immediate consequence the following inequality:

$$\|f_A(z)\|_{B(\ell_2)} \leq \|A\|_{\mathcal{B}(D,\ell_2)} \log \frac{1 + |z|}{1 - |z|}$$

for all $|z| \geq \frac{1}{2}$, which implies that the point evaluation of a matrix A, that is the linear operator $\Delta_z : \mathcal{B}(D, \ell_2) \to B(\ell_2)$, given by $\Delta_z(A) = f_A(z)$, is bounded for all $z \in D$. This, in its turn, implies that if a sequence A^n of matrices converges in the Bloch norm, then $\Delta_z(A^n)$ does so locally uniformly with respect to z.

In particular, we can use the inequality above to prove the completeness of $\mathcal{B}(D, \ell_2)$.

Proposition 7.14. *The space of Bloch matrices $\mathcal{B}(D, \ell_2)$ is a Banach space.*

Proof. Denoting the dilations of f_A by $(f_A)_r(z) = f_A(rz) = f_{A*C(r)}(z)$, we have that

$$\|A * C(r)\|_{\mathcal{B}(D,\ell_2)} = \sup_{z \in D}(1 - |z|^2)\|(f_A)'_r(z)\|_{B(\ell_2)} + \|(f_A)_r(0)\|_{B(\ell_2)} =$$

$$r \sup_{z \in D}(1 - |z|^2)\|(f_A)'(rz)\|_{B(\ell_2)} + \|f_A(0)\|_{B(\ell_2)},$$

which increases to $\|A\|_{\mathcal{B}(D,\ell_2)}$ as r increases to 1. Using this fact, we can now show that $\mathcal{B}(D, \ell_2)$ is complete.

Let $\{A^n\}$ be a Cauchy sequence in $\mathcal{B}(D, \ell_2)$. By $(*)$, this implies that $\{f_{A^n}\}$ is a uniform Cauchy sequence on each compact subset of D, and, hence, it converges locally uniformly to some vector-valued analytic function f_A. It remains to show that $\|A^n - A\|_{\mathcal{B}(D,\ell_2)} \to 0$. Given $\epsilon > 0$, choose N such that $\|A^n - A^m\|_{\mathcal{B}(D,\ell_2)} < \epsilon$ when $n, m \geq N$. Then, for $r < 1$,

$$\|A^n * C(r) - A * C(r)\|_{\mathcal{B}(D,\ell_2)} \leq \|A^n * C(r) - A^m * C(r)\|_{\mathcal{B}(D,\ell_2)} +$$

$$\|A^m * C(r) - A * C(r)\|_{\mathcal{B}(D,\ell_2)} \leq \|A^n - A^m\|_{\mathcal{B}(D,\ell_2)}$$

$$+ \|A^m * C(r) - A * C(r)\|_{\mathcal{B}(D,\ell_2)} < \epsilon + \|A^m * C(r) - A * C(r)\|_{\mathcal{B}(D,\ell_2)}.$$

Observe that the last term approaches 0 as $m \to \infty$, since $(A^m * C(r))'$ converges to $(A * C(r))'$ uniformly on D. Thus

$$\|A^n * C(r) - A * C(r)\|_{\mathcal{B}(D,\ell_2)} \leq 2\epsilon \text{ for } n \geq N \text{ and all } r < 1.$$

Finally, we let $r \to 1$ to arrive at the desired conclusion. The proof is complete. $\qquad\square$

7.2 Matrix version of little Bloch space

Now we introduce another space of matrices, the so-called *little Bloch space of matrices*.

Definition 7.15. *The space $\mathcal{B}_0(D, \ell_2)$ is the space of all upper triangular infinite matrices A such that $\lim_{r \to 1^-} (1 - r^2)\|(A * C(r))'\|_{B(\ell_2)} = 0$, where $C(r)$ is the Toeplitz matrix associated with the Cauchy kernel.*

Clearly $\mathcal{B}_0(D, \ell_2)$ is a closed subspace of $\mathcal{B}(D, \ell_2)$ if the former space is endowed with the norm of $\mathcal{B}(D, \ell_2)$.

Let $A \in \mathcal{B}(D, \ell_2)$ and $A_r(s) = A(rs) = A(r) * P(s)$ for all $0 \leq r < 1$ and $0 \leq s < 1$, where $P(s)$ is the Toeplitz matrix associated to the Poisson kernel, that is

$$P(s) = \begin{pmatrix} 1 & s & s^2 & s^3 & \cdots \\ s & 1 & s & s^2 & \ddots \\ s^2 & s & 1 & s & \ddots \\ s^3 & s^2 & s & 1 & \ddots \\ \vdots & \ddots & \ddots & \ddots & \ddots \end{pmatrix}.$$

Then it follows that A_r is a matrix belonging to $\mathcal{B}_0(D, \ell_2)$, for all $0 \le r < 1$, since

$$\lim_{s \to 1}(1 - s^2)||A'_r(s)||_{B(\ell_2)} \le \lim_{s \to 1} \frac{(1 - s^2)}{1 - r^2 s^2}||A||_{\mathcal{B}(D,\ell_2)} \cdot r \int_{-\pi}^{\pi} \frac{d\theta}{|1 - se^{i\theta}|}$$

$$\sim ||A||_{\mathcal{B}(D,\ell_2)} \cdot \frac{r}{1 - r^2} \lim_{s \to 1}(1 - s^2) \log \frac{1}{1 - s^2} = 0.$$

Theorem 7.16. *Let* $A \in \mathcal{B}(D, \ell_2)$. *Then* $A \in \mathcal{B}_0(D, \ell_2)$ *if and only if* $\lim_{r \to 1^-} ||A_r - A||_{\mathcal{B}(D,\ell_2)} = 0.$

Proof.   By the remark above it follows that $A_r \in \mathcal{B}_0(D, \ell_2)$ and we use the obvious fact that $\mathcal{B}_0(D, \ell_2)$ is a closed subspace of $\mathcal{B}(D, \ell_2)$ in order to conclude that the condition is sufficient.

Conversely, let $A \in \mathcal{B}_0(D, \ell_2)$. Then $\forall \epsilon > 0$ there is $0 < \delta < 1$ such that $(1 - s^2)||A'(s)||_{B(\ell_2)} < \epsilon \; \forall \delta^2 < s < 1$. We note that

$$||A_r - A||_{\mathcal{B}(D,\ell_2)} = \sup_{0 \le s < 1} (1 - s^2)||A'_r(s) - A'(s)||_{B(\ell_2)}$$

$$\le \sup_{\delta < s < 1} (1 - s^2)||A'_r(s) - A'(s)||_{B(\ell_2)}$$

$$+ \sup_{0 \le s \le \delta} (1 - s^2)||A'_r(s) - A'(s)||_{B(\ell_2)}.$$

For $\delta < r < 1$, the first term is smaller than $(1 - r^2 s^2)||A'(rs)||_{B(\ell_2)} + (1 - s^2)||A'(s)||_{B(\ell_2)} < 2\epsilon$.

The second term converges to 0 whenever $r \to 1^-$. Indeed, for $0 \le s \le \delta < \delta' < 1$, letting $u = \frac{s}{\delta'}$, we get that

$$||A'_r(s) - A'(s)||_{B(\ell_2)} = ||A'(s) * \sum_{k=0}^{\infty}(r^k - 1)E_k||_{B(\ell_2)}$$

$$\le ||A'(u)||_{B(\ell_2)} \cdot ||\sum_{k=0}^{\infty}(r^k - 1)(\delta')^{k-1}E_k||_{M(\ell_2)}$$

$$= ||A'(u)||_{B(\ell_2)} \cdot ||\sum_{k=0}^{\infty}(r^k - 1)\delta'^{k-1}e^{ik\theta}||_{M(\mathbb{T})}$$

$$\le ||A'(u)||_{B(\ell_2)} \cdot \int_{-\pi}^{\pi} \frac{|1 - \delta'e^{i\theta} - 1 + r\delta'e^{i\theta}|}{|1 - r\delta'e^{i\theta}| \cdot |1 - \delta'e^{i\theta}|} \frac{d\theta}{2\pi}$$

$$\leq ||A'(\frac{s}{\delta'})||_{B(\ell_2)} \cdot \frac{(1-r)\delta'}{(1-r\delta')(1-\delta')}.$$

We recall that E is the Toeplitz matrix having all its entries equal to 1. Therefore,

$$\sup_{0 \leq s \leq \delta} (1-s^2)||A'(rs) - A'(s)||_{B(\ell_2)} \leq$$

$$\sup_{0 \leq s \leq \delta < \delta'} \left[(1 - (\frac{s}{\delta'})^2)||A'(\frac{s}{\delta})||_{B(\ell_2)} \cdot \frac{1-s^2}{1 - \frac{s^2}{\delta'^2}} \right] \cdot \frac{\delta'(1-r)}{(1-\delta')(1-r\delta')} \leq$$

$$||A||_{B(D,\ell_2)} \cdot \frac{1-\delta^2}{1 - \frac{\delta^2}{\delta'^2}} \cdot \frac{\delta'(1-r)}{(1-\delta')(1-r\delta')},$$

for all $t \geq 0$.

Consequently

$$\lim_{r \to 1^-} \sup_{s \leq \delta}(1-s^2)||A'_r(s) - A'(s)||_{B(\ell_2)} = 0.$$

The proof is complete. $\qquad\qquad\qquad\qquad\qquad\qquad\qquad\qquad\qquad\quad$ □

Corollary 7.17. $\mathcal{B}_0(D, \ell_2)$ *is the closure of all matrices of finite band type in the Bloch norm. In particular, this implies that $\mathcal{B}_0(D, \ell_2)$ is a separable space.*

Proof. Let $A \in \mathcal{B}_0(D, \ell_2)$ and $A^n = \sum_{k=0}^n A_k$.. Then, by Theorem 7.16, we have that $\forall \epsilon > 0$ there is a $r_0 < 1$ such that $||A_r - A||_{B(D,\ell_2)} < \epsilon/2$.

We note that $r \to A(r)$ for $r \in [0,1)$ is a continuous $B(\ell_2)$-valued function on $[0, s]$ for $s < 1$.

Indeed, let $0 < s_n \leq s_0 < 1$ and $s_n \to s_0$. Then

$$||A(s_n) - A(s_0)||_{B(\ell_2)} = ||[C(\frac{s_n}{s'}) - C(\frac{s_0}{s'})] * A(s')||_{B(\ell_2)}$$

$$\leq ||A(s')||_{B(\ell_2)} \cdot ||C(\frac{s_n}{s'}) - C(\frac{s_0}{s'})||_{M(\ell_2)},$$

where $s_n \to s_0 < s' < 1$. Hence, by putting $\delta = \frac{s_0}{s'} < 1$ and reasoning as in the proof of Theorem 7.16, we get that

$$||A(s_n) - A(s_0)||_{B(\ell_2)} \leq ||A(s')||_{B(\ell_2)} \cdot ||\sum_{k=1}^{\infty} \delta^k((\frac{s_n}{s_0})^k - 1)e^{ik\theta}||_{M(\mathbb{T})} \to 0.$$

Moreover, for a fixed $r_0 < 1$ we have that

$$\sup_{0 \leq s \leq 1} ||A_{r_0}(s)||_{B(\ell_2)} := M(r_0) = ||A(r_0)||_{B(\ell_2)} < \infty$$

for all analytic matrices.

Thus, for $r_0 < r' < 1$ and by using the notation

$$C^n\left(\frac{r_0}{r'}\right) = \sum_{k=0}^{n} C_k\left(\frac{r_0}{r'}\right),$$

we find that

$$\|A_{r_0}(\cdot) - (A_{r_0})^n(\cdot)\|_{L^\infty(D,\ell_2)} = \operatorname*{ess\,sup}_{s<1}\|(A - A^n)(r_0 s)\|_{B(\ell_2)}$$

$$= \|(A - A^n)(r_0)\|_{B(\ell_2)}$$

$$= \|[C(\frac{r_0}{r'}) - C^n(\frac{r_0}{r'})] * A(r')\|_{B(\ell_2)} \le \|C(\frac{r_0}{r'}) - C^n(\frac{r_0}{r'})\|_{M(\ell_2)} \cdot \|A(r')\|_{B(\ell_2)}$$

$$= \|\sum_{k+1}^{\infty}(\frac{r_0}{r'})^k e^{ik\theta}\|_{M(\mathbb{T})} \cdot \|A(r')\|_{B(\ell_2)} \to 0.$$

Consequently,

$$\|A_{r_0}(\cdot) - (A_{r_0})^n(\cdot)\|_{L^\infty(D,\ell_2)} \to 0$$

and

$$\|A(\cdot) - (A_{r_0})^n(\cdot)\|_{\mathcal{B}(D,\ell_2)} \le \epsilon,$$

whenever $r_0 < 1$ is fixed as before and n is sufficiently large. The proof is complete. □

The next theorem expresses a natural relation between the Bergman projection and the Bloch spaces.

Theorem 7.18. *Let $A \in \mathcal{B}(D, \ell_2)$. The following assertions are equivalent:*

1) $A \in \mathcal{B}_0(D, \ell_2)$.

2) There is a continuous function $r \to B(r)$, defined on $[0, 1]$ and $B(\ell_2)$-valued, such that $P(B(\cdot))(r) = A(r)$.

3) There is a function $r \to B(r)$, which is a continuous $B(\ell_2)$-valued function such that $\lim_{r \to 1} B(r) = 0$ and satisfying that $P(B(\cdot))(r) = A(r)$.

Proof. To prove that 1) implies 3), we take $A \in \mathcal{B}_0(D, \ell_2)$. We define $A_1(r) = \sum_{k=2}^{\infty} A_k r^k$, with $r < 1$. Thus $A_1'(r) = \sum_{k=2}^{\infty} k A_k r^{k-1}$ and $A(r) = A_0 + A_1 r + A_1(r)$. We take now

$$B_2(r) = (1 - r^2)T * P(r) * A_1'(r),$$

where $P(r)$ is the Toeplitz matrix associated to the Poisson kernel and $T = (t_{ij})_{i,j}$ is a Schur multiplier, which will be defined later on.

Thus, by the definition of Bergman projection P, we get that

$$[PB_2(\cdot)](r)(i,j)$$

$$= \begin{cases} t_{ij}a_{ij}(j-i+1)(j-i)r^{j-i}\dfrac{1}{[\frac{3}{2}(j-i+1)]^{\frac{3(j-i)+1}{2}}} & \text{for } j-i \geq 2, \\ 0 & \text{if } j-i < 2. \end{cases}$$

Consequently, by taking

$$t_{ij} = \begin{cases} \dfrac{3[3(j-i)+1]}{4(j-i)} & \text{for all } j \neq i, \ i,j \geq 1 \\ \dfrac{9}{4} & \text{if } j = i, i \geq 1, \end{cases}$$

it follows that T is a Schur multiplier, and $P[B_2(\cdot)] = A_1(\cdot)$. Let now

$$B(r) = 2(1-r^2)A_0 + 3(1-r^2)rA_1 + B_2(r).$$

It is clear that $[PB(\cdot)](r) = A(r)$. But, since $A \in \mathcal{B}_0(D, \ell_2)$, it follows that $B_2(r)$, and, consequently, $B(r)$, is a continuous $B(\ell_2)$-valued function, and $\lim_{r \to 1} B(r) = 0$. Thus 3) holds and we have proved that 1) implies 3).

It is obvious that 3) implies 2).

It remains to prove that 2) implies 1). Let 2) hold and choose $B(r) \in B(\ell_2)$ such that

$$[PB(\cdot)](r) = A(r) \text{ for } r \in [0,1].$$

Assume that $r \to B(r)$ is a continuous $B(\ell_2)$-function on $[0,1]$ and let $M = \sup_{0 \leq r \leq 1} \|B(r)\|_{B(\ell_2)} < \infty$.

Let $0 \leq r_0 < 1$ be fixed and consider $A_{r_0}(r)$ given by the formula

$$A_{r_0}(r)(i,j) = \begin{cases} (j-i+1)(rr_0)^{j-i}(2\int_0^1 b_{ij}(s)s^{j-i+1}ds) & \text{if } j-i \geq 0, \\ 0 & \text{otherwise.} \end{cases}$$

Consequently, according to Theorem 7.3, we find that

$$A_{r_0} = P[P(r_0) * B(\cdot)] = P[B_{r_0}(\cdot)] \in \mathcal{B}(D, \ell_2),$$

where $P(r_0)$ is the Toeplitz matrix associated to the Poisson kernel.

Let $\mathcal{C}(\overline{D}, \ell_2)$ denote the space of all continuous $\mathcal{B}(D, \ell_2)$-valued functions defined on $[0,1]$. Next we prove that the function $s \to P(P(r_0) * B(s))$ belongs to $\mathcal{C}(\overline{D}, \ell_2)$ if B is a continuous $B(\ell_2)$-valued function such that

$$\lim_{r \to 1} \sup_{s \in [0,1]} \|A_r(s) - A(s)\|_{B(\ell_2)} = 0. \tag{7.4}$$

This, in its turn, implies that $\lim_{r \to 1} A_r = A$ in $\mathcal{B}(D, \ell_2)$. Therefore, by the Theorem 7.16, it follows that $A \in \mathcal{B}_0(D, \ell_2)$.

Let $s, s_0 \in [0,1]$. Then

$$||P(r_0)*[B(s)-B(s_0)]||_{B(\ell_2)} \leq ||\sum_{k\in\mathbb{Z}} r_0^{|k|} e^{ik\theta}||_{M(\mathbb{T})} \cdot ||B(s)-B(s_0)||_{B(\ell_2)} \to 0$$

for $s \to s_0$ and $B(s)$ is a continuous function on $[0,1]$. Here we have used Bennett's Theorem and the fact that

$$||\sum_{k\in\mathbb{Z}} r^{|k|} e^{ik\theta}||_{M(\mathbb{T})} = ||\frac{1-r^2}{|1-re^{i\theta}|^2}||_{M(\mathbb{T})} \leq 1.$$

Thus, the function $s \to P[P(r_0)*B(s)]$ belongs to $\mathcal{C}(\overline{D}, \ell^2)$.

Hence, it only remains to prove that (7.4) holds. In fact,

$$||P(r_0)*B(s)-B(s)||_{B(\ell_2)} \leq ||\sum_{k\in\mathbb{Z}}(r_0^{|k|}-1)e^{ik\theta}||_{M(\mathbb{T})} \cdot ||B(s)||_{B(\ell_2)}$$

$$\leq M \cdot ||\sum_{k\in\mathbb{Z}}(r^{|k|}-1)e^{ik\theta}||_{M(\mathbb{T})} \text{ for all } s \in [0,1].$$

Denoting by $\mu_r(\theta)$ the measure $\sum_{k\in\mathbb{Z}}(r^{|k|}-1)e^{ik\theta}$, then , for a trigonometric polynomial $\phi(\theta) = \sum_{n=-m}^{m} a_n e^{in\theta}$, we have that

$$\mu_r(\phi) = \sum_{n=-m}^{m} (r^{|n|}-1)\overline{a_n} \quad \text{and} \quad |\mu_r(\phi)| \leq |\overline{\phi(r) - \phi(1)}| \leq 2||\phi||,$$

where $\phi(r)$ is the value of the Poisson extension of ϕ in the point r.

Consequently μ_r is a measure with the norm less than 2. But $\lim_{r\to 1} \mu_r(\phi) = 0$ for all trigonometric polynomes ϕ. Thus, w^*- $\lim_{r\to 1} \mu_r = 0$ in $M(\mathbb{T})$ and then it is clear that $\lim_{r\to 1} ||\mu_r|| = 0$ and, according to Theorem 7.3, the relation (7.4) is proved. Thus, also the implication 2) $\Rightarrow$ 1) is proved and the proof is complete. $\qquad\square$

Now, using the notation which preceeds Theorem 6.12, we get:

Theorem 7.19. P_2 *is a continuous operator (precisely a continuous projection) from* $L^1(D, \ell_2)$ *onto* $L_a^1(D, \ell_2)$.

Proof. By Theorem 1.2 the topological dual of $L^1(D, \ell_2))$ is $L^\infty(D, \ell_2)$ with respect to the duality pair:

$$< A(\cdot), B(\cdot) > := \int_0^1 tr\,(A(s)[B(s)]^*)2s\,ds,$$

where $A(\cdot) \in L^\infty(D, \ell_2)$, $B(\cdot) \in L^1(D, \ell_2)$. By using a duality argument it is sufficient to prove that $P_2^* : L^\infty(D, \ell_2) \to L^\infty(D, \ell_2)$ is bounded.

We are looking for the adjoint P_2^* of P_2 :

$$< P_2^* A(\cdot), B(\cdot) > = 2 \int_0^1 \sum_{i=1}^\infty \sum_{j=1}^\infty (P_2^* A(\cdot))(r)(i,j)\overline{b_{ij}(r)}r dr$$

$$= \sum_{i=1}^\infty \sum_{j=1}^\infty \int_0^1 (P_2^* A(\cdot))(r)(i,j)\overline{b_{ij}(r)}(2r)dr.$$

On the other hand it yields that

$$< P_2^* A(\cdot), B(\cdot) > = < A(\cdot), P_2 B(\cdot) > = \int_0^1 \operatorname{tr} A(r)(P_2 B)^*(r)(2r dr)$$

$$= \sum_{i=1}^\infty \sum_{j=1}^\infty \int_0^1 A(r)(i,j)\overline{(P_2 B)(r)}(i,j)(2r dr)$$

$$= \sum_{i=1}^\infty \sum_{j=i}^\infty \frac{\Gamma(j-i+4)}{(j-i)!\Gamma(3)} \left(\int_0^1 [A(s)](i,j)s^{j-i}(2sds) \right) \times$$

$$\left(\int_0^1 \overline{b_{ij}(s)}s^{j-i}(1-s^2)^2(2sds) \right).$$

Now we consider $\{I_k\}$, a sequence of intervals such that

$$\lim_{k\to\infty} \mu(I_k) = 0, d\mu = 2sds \text{ and } r \in I_k.$$

For every k, we take $B(s)(i,j) = \chi_{I_k}(s)/(\mu(I_k))$ and $B(s)(l,k) = 0$, $(l,k) \neq (i,j)$ for every $(i,j) \in \mathbb{N} \times \mathbb{N}$.

By Lebesgue's differentiation theorem we have that

$$(P_2^* A(\cdot))(r)(i,j) = \begin{cases} \frac{\Gamma(j-i+4)}{(j-i)!\Gamma(3)} r^{j-i}(1-r^2)^2 \int_0^1 A(s)(i,j)s^{j-i}(2sds) & \text{if } j \geq i, \\ 0 & \text{if } j < i, \end{cases}$$

a.e. for all $r \in [0,1)$.

We will now prove that $P_2^* : L^\infty(D,\ell_2) \to L^\infty(D,\ell_2)$ is a bounded operator. In order to prove that we first note that

$$||A(r)||^2_{L^\infty(D,\ell_2)} = \operatorname*{ess\,sup}_{0 \leq r < 1} ||A(r)||^2_{B(\ell_2)}$$

$$= \operatorname*{ess\,sup}_{0 \leq r < 1} \sup_{\sum_{j=1}^\infty |h_j|^2 \leq 1} \sum_{i=1}^\infty |\sum_{j=1}^\infty a_{ij}(r)h_j|^2.$$

Consequently, because $L^1[0,1]$ has cotype 2, there is a constant $K > 0$, such that

$$\|P_2^* A(\cdot)\|_{L^\infty(D,\ell_2)}^2 =$$

$$\|\sup_{\|h\|_{\ell_2}\leq 1} \sum_{i=1}^\infty |\sum_{j=1}^\infty \int_0^1 h_j r^{j-i} \frac{\Gamma(j-i+4)}{(j-i)!\Gamma(3)}(1-r^2)^2 a_{ij}(s)s^{j-i}(2sds)|^2\|_{L^\infty} =$$

$$\|\sup_{\|h\|_{\ell_2}\leq 1} (1-r^2)^4 \sum_{i=1}^\infty |\int_0^1 (\sum_{j=i}^\infty a_{ij}(s)((rs)^{j-i}\frac{\Gamma(j-i+4)}{(j-i)!\Gamma(3)})h_j)(2sds)|^2\|_{L^\infty} \leq$$

$$K\|(1-r^2)^4 \sup_{\|h\|_{\ell_2}\leq 1} [\int_0^1 (\sum_{i=1}^\infty |\sum_{j=i}^\infty a_{ij}(s)[(rs)^{j-i}\frac{(j-i+3)!}{(j-i)!}h_j|^2)^{1/2}(sds)]^2\|.$$

Since the Toeplitz matrix $C(rs) = ((c_{ij})(rs)_{i,j=1}^\infty)$, where

$$c_{ij}(rs) := c_{j-i}(rs) = \begin{cases} (rs)^{j-i}(j-i+3)(j-i+2)(j-i+1) & \text{if } j \geq i \\ 0 & \text{if } j < i, \end{cases}$$

is a Schur multiplier with

$$\|C(rs)\|_{L^1(\mathbb{T})} = \|\frac{6}{(1-rse^{i\theta})^4}\|_{L^1(\mathbb{T})} = 6\sum_{n=0}^\infty (n+1)^2(rs)^{2n},$$

we get that

$$\sup_{\sum_{j=1}^\infty |h_j|^2 \leq 1} \left(\sum_{i=1}^\infty |\sum_{j=i}^\infty a_{ij}(s)(rs)^{j-i}(j-i+3)(j-i+2)(j-i+1)h_j|^2\right)^{1/2}$$

$$= \|A(s) * C(rs)\|_{B(\ell_2)} \leq 6\|A(s)\|_{B(\ell_2)} \cdot \sum_{n=0}^\infty (n+1)^2(rs)^{2n}.$$

Hence,

$$\|P_2^* A(\cdot)\|_{L^\infty(D,\ell_2)}^2$$

$$\leq K \operatorname*{ess\,sup}_{r<1}(1-r^2)^4 \left[\int_0^1 3\|A(s)\|_{B(\ell_2)} \cdot \sum_{n=0}^\infty (n+1)^2 r^{2n}s^{2n}(2sds)\right]^\iota$$

$$\leq 9K \cdot \operatorname*{ess\,sup}_{r<1}(1-r^2)^4\|A(\cdot)\|_{L^\infty(D,\ell_2)}^2 \left(\sum_{n=0}^\infty (n+1)r^{2n}\right)^2 \sim$$

$$\operatorname*{ess\,sup}_{r<1}(1-r^2)^4\frac{1}{(1-r^2)^4}\|A(\cdot)\|_{L^\infty(D,\ell_2)}^2 \sim \|A(\cdot)\|_{L^\infty(D,\ell_2)}^2,$$

which shows in turn that $P_2^* : L^\infty(D,\ell_2) \to L^\infty(D,\ell_2)$ is bounded. The proof is complete. □

We denote by $\mathcal{C}_0(D, \ell_2)$ the space of all continuous $B(\ell_2)$-valued functions $B(r)$ on $[0,1]$ such that $\lim_{r \to 1} B(r) = 0$ in the norm of $B(\ell_2)$.

Lemma 7.20. *Let* $V = (P_2)^*$, *that is*

$$(P_2)^*(A(r))(i,j) =$$

$$\begin{cases} \frac{(j-i+3)(j-i+2)(j-i+1)}{2} r^{j-i}(1-r^2)^2 \int_0^1 a_{ij}(s)s^{j-i}(2s\,ds) & \text{if } j-i \geq 0, \\ 0 & \text{otherwise.} \end{cases}$$

Then V *is an isomorphic embedding of* $\mathcal{B}_0(D, \ell_2)$ *in* $\mathcal{C}_0(D, \ell_2)$.

Proof. According to Theorem 7.18, for $B \in \mathcal{B}_0(D, \ell_2)$ we can find some $A(\cdot) \in \mathcal{C}_0(D, \ell_2)$ such that $[PA(\cdot)](r) = B(r)$. Clearly, we have that

$$P_2^*B = P_2^*PA = T^1 * (P_2^*A_1),$$

where $A_1(r) = T * A(r)$, for $T = (t_{j-i})_{i,j}$ with

$$t_{j-i} = \begin{cases} \frac{2(j-i+1)}{j-i+2} & \text{for } j-i \neq -2, \\ 0 & \text{otherwise.} \end{cases}$$

T is a Schur multiplier and the same is true for $T^1 = (t^1_{j-i})_{i,j}$, where

$$t^1_{j-i} = \begin{cases} \frac{j-i+2}{2(j-i+1)} & \text{for } j-i \neq -1, \\ 0 & \text{otherwise.} \end{cases}$$

Thus $\|A_1(r)\|_{B(\ell_2)} \sim \|A(r)\|_{B(\ell_2)}$ for all $r \in [0,1]$.

Hence, we obtain that

$$\|P_2^*A_1(r)\|^2_{B(\ell_2)}$$

$$= \sup_{\|h\|_{\ell_2} \leq 1} \sum_{i=1}^{\infty} |\sum_{j=i}^{\infty} h_j r^{j-i} \frac{\Gamma(j-i+4)}{(j-i)!\Gamma(3)} (1-r^2)^2 \int_0^1 a^1_{ij}(s)s^{j-i}(2s\,ds)|^2 \leq$$

$$K \sup_{\|h\|_{\ell_2} \leq 1} (1-r^2)^4 [\int_0^1 (\sum_{j=1}^{\infty} |\sum_{j=i}^{\infty} a^1_{ij}(s)(rs)^{j-i} \frac{\Gamma(j-i+4)}{2(j-i)!} h_j|^2)^{1/2} (2s\,ds)]^2.$$

The Toeplitz matrix $C(r,s) = (c_{j-i}(r,s))_{i,j}$, where

$$c_{j-i}(r,s) = \begin{cases} (rs)^{j-i}(j-i+3)(j-i+1)^2 & \text{if } j \geq i \\ 0 & \text{otherwise,} \end{cases}$$

is a Schur multiplier, since

$$\sum_{k=0}^{\infty} (rs)^k (k+3)(k+1)^2 e^{ik\theta} \sim \frac{1}{(1-rse^{i\theta})^4}$$

and

$$\int_{-\pi}^{\pi} \frac{1}{|1-rse^{i\theta}|^4} \frac{d\theta}{2\pi} \sim \sum_{n=0}^{\infty} n^2 (rs)^{2n}.$$

Therefore, we have that

$$\|P_2^* A_1(r)\|_{B(\ell_2)} \leq C(1-r^2)^2 \int_0^1 \|A(s)\|_{B(\ell_2)} \cdot (\sum_{n=0}^{\infty} n^2 (rs)^{2n})(2s\,ds).$$

Since $\lim_{s\to 1} \|A(s)\|_{B(\ell_2)} = 0$, for all $\epsilon > 0$, there is $\delta > 0$ such that $\|A(s)\| < \epsilon$ for all $1 \geq s \geq \delta$ and, consequently,

$$\|P_2^* A_1(r)\|_{B(\ell^2)} \leq C(\epsilon + \|A(\cdot)\|_{\mathcal{C}(D,\ell^2)} \cdot \frac{(1-r^2)^2 \delta^2}{(1-r^2\delta^2)^2}).$$

It follows that $\lim_{r\to 1} \|P_2^* A_1(r)\|_{B(\ell^2)} = 0$, and, since T^1 is a Schur multiplier, it follows that $P_2^* B \in \mathcal{C}_0(D, \ell^2)$ and $\|P_2^* B\|_{B(\ell^2)} \leq C \|A(\cdot)\|_{\mathcal{C}(D,\ell^2)}$.

Moreover, in view of the proof of Theorem 7.18, we can find an $A(\cdot) \in \mathcal{C}_0(D, \ell_2)$ such that

$$\|A(\cdot)\|_{\mathcal{C}_0(D,\ell_2)} \leq C(\|B_0\|_{B(\ell_2)} + \|B(\cdot)\|_{\mathcal{B}(D,\ell_2)}),$$

where $C > 0$ is an absolute constant. By now also using the arguments in the proof of Theorem 7.19 it follows that $P_2^* : \mathcal{B}_0(D, \ell_2) \to \mathcal{C}_0(D, \ell_2)$ is bounded.

On the other hand, if $A \in \mathcal{B}_0(D, \ell_2)$, then, since $A(r)$ is an analytic matrix, it is obvious that $A(r) = [PA(\cdot)](r) = (P[P_2^* A(\cdot)])(r)$ for all $r \in [0,1)$. Thus, by using Theorem 7.19, we conclude that there exists a constant $C > 0$ such that $\|A(\cdot)\|_{\mathcal{B}(D,\ell_2)} \leq C \|P_2^* A(\cdot)\|_{\mathcal{C}(D,\ell_2)}$, which implies that $P_2^* : \mathcal{B}_0(D, \ell_2) \to \mathcal{C}_0(D, \ell_2))$ is an isomorphic embedding. The proof is complete.   □

From now on we identify $\mathcal{B}_0(D, \ell^2)$ with the space $\widetilde{\mathcal{B}}_0(D, \ell^2)$ of all analytic matrices $A * C(r)$, for $A \in \mathcal{B}_0(D, \ell^2)$.

We introduce $\mathcal{B}_{0,c}(D, \ell_2)$ as the closed Banach subspace of $\mathcal{B}_0(D, \ell_2)$ consisting of all upper triangular matrices whose diagonals are compact operators. We are now ready to prove that this space $\mathcal{B}_{0,c}(D, \ell_2)$ is in fact the predual of the Bergman-Schatten space. More exactly, our last main result in this Section is the following duality result:

Theorem 7.21. *It yields that* $\mathcal{B}_{0,c}(D, \ell_2)^* = L_a^1(D, \ell_2)$ *with respect to the usual duality, whenever* $\mathcal{B}_0(D, \ell_2)$ *has the norm induced by* $\mathcal{B}(D, \ell_2)$.

Proof. Let $A \in L_a^1(D, \ell_2)$. Then $B \to \int_0^1 tr[B(s)A^*(s)](2sds)$ defines a linear and bounded functional on $\mathcal{B}_{0,c}(D, \ell_2)$ due to Theorem 7.11. Conversely, let us assume that F is a bounded linear functional on $\mathcal{B}_{0,c}(D, \ell_2)$. Then we shall show that there is a matrix C from $L_a^1(D, \ell_2)$ such that

$$F(B) = \int_0^1 tr[B(r)C^*(r)](2rdr), \tag{7.5}$$

for B from a dense subset of $\mathcal{B}_0(D, \ell_2)$.

In what follows we identify the matrices $A \in \mathcal{B}_0(D, \ell_2)$ with the function $r \to A * C(r)$. In particular, we identify $\sum_{k=0}^n r^k A_k$ with $\sum_{k=0}^n A_k$.

By Lemma 7.20 it follows that $P_2^* : \mathcal{B}_0(D, \ell_2) \to \mathcal{C}_0(D, \ell_2)$, is an isomorphic embedding. Thus $X = P_2^* (\mathcal{B}_{0,c}(D, \ell_2))$ is a closed subspace in $\mathcal{C}_0(D, C_\infty)$ and $F \circ (P_2^*)^{-1} : X \to \mathbb{C}$, is a bounded linear functional on X, where $\mathcal{C}_0(D, C_\infty)$ is the subset in $\mathcal{C}_0(D, \ell_2)$ whose elements are C_∞-valued functions. By Hahn-Banach theorem $F \circ (P_2^*)^{-1}$ can be extended to a bounded linear functional on $\mathcal{C}_0(D, C_\infty)$.

Let $\Phi : \mathcal{C}_0(D, C_\infty) \to \mathbb{C}$ denote this functional. It follows that $\mathcal{C}_0(D, C_\infty) = \mathcal{C}_0[0,1] \hat{\otimes}_\epsilon C_\infty$, Thus, Φ is a bilinear integral map, i.e., there is a bounded Borel measure μ on $[0,1] \times U_{C_1}$, where U_{C_1} is the unit ball of the space C_1 with the topology $\sigma(C_1, C_\infty)$, such that

$$\Phi(f \otimes A) = \int_{[0,1] \times U_{C_1}} f(r)tr(AB^*)d\mu(r, B) \quad \forall f \in \mathcal{C}_0[0,1] \text{ and } A \in C_\infty.$$

Thus, for the matrix $\sum_{k=0}^n A_k \in \mathcal{B}_{0,c}(D, \ell_2)$, identified with the analytic matrix $\sum_{k=0}^n A_k r^k$, we have that

$$F(\sum_{k=0}^n A_k) = F(\sum_{k=0}^n r^k A_k) = [F \circ (P_2^*)^{-1}][P_2^*(\sum_{k=0}^n r^k A_k)]$$

$$= \Phi(\sum_{k=0}^n \frac{(k+3)(k+2)}{2} r^k (1-r^2)^2 A_k)$$

$$= \int_{[0,1] \times U_{C_1}} \sum_{k=0}^n tr[(\frac{(k+3)(k+2)}{2} r^k (1-r^2)^2 A_k) B^*] d\mu(r, B)$$

$$:= < \mu(r, B), tr(\sum_{k=0}^n \frac{(k+3)(k+2)}{2} r^k A_k) B^* (1-r^2)^2 > .$$

On the other hand, we wish to prove that

$$F(\sum_{k=0}^{n} A_k) = \int_0^1 \text{tr}(\sum_{k=0}^{n} s^k A_k)(C(s)^*)(2sds) = \int_0^1 \text{tr}(\sum_{k=0}^{n} s^{2k} A_k C_k^*)(2sds)$$

$$= \sum_{k=0}^{n} \text{tr} A_k (\frac{C_k^*}{k+1}).$$

Now, letting $A = e_{i,i+k}$, where $e_{i,i+k}$ is the matrix having 1 as the single nonzero entry on the ith-row and the $(i+k)$th-column, for $i \geq 1$ and $j \geq 0$, we find that

$$C_k = \; < \overline{\mu}(r, B), \frac{(k+1)(k+2)(k+3)}{2} r^k (1 - r^2)^2 B_k >, \; k = 0, 1, 2, \ldots.$$

Therefore, denoting $[0,1] \times U_{C_1}$ by V, we have

$$\int_0^1 ||C(s)||_{C_1}(2sds)$$

$$= \int_0^1 || \int_V \sum_{k=0}^{n} \frac{(k+3)!}{2\,k!}(sr)^k (1 - r^2)^2 B_k d\overline{\mu}(r, B)||_{C_1}(2sds) \leq$$

$$\int_V [\int_0^1 || \sum_{k=0}^{n} \frac{(k+3)!}{2\,k!}(rs)^k (1 - r^2)^2 B_k ||_{C_1})(2sds)] d|\mu|(r, B) \leq$$

$$\int_V [\int_0^1 || \sum_{k=0}^{n} \frac{(k+3)!}{2\,k!}(rs)^k (1 - r^2)^2 e^{ik(\cdot)}||_{L^1(\mathbb{T})} ||B||_{C_1})(2sds)] d|\mu|(r, B)$$

$$\sim \int_V [\int_0^1 \int_0^{2\pi} \frac{(1 - r^2)^2}{|1 - rse^{i\theta}|^4} \frac{d\theta}{2\pi}(2sds)] d|\mu|(r, B)$$

$$= \int_V \int_0^1 (1 - r^2)^2 \sum_{k=0}^{\infty} (n + 1)^2 (sr)^{2n}(2sds) d|\mu|(r, B)$$

$$= \int_V (1 - r^2)^2 \sum_{n=0}^{\infty} (n + 1) r^{2n} d|\mu|(r, B) \leq ||\mu|| < \infty.$$

Consequently, $C \in L_a^1(D, \ell_2)$ and we get the relation (7.5), using the fact that the set of all matrices $\sum_{k=0}^{n} A_k$ is dense in $\mathcal{B}_{0,\,c}(D, \ell_2)$. The proof is complete. $\qquad\square$

Notes

The idea behind our study of matrix versions of different kinds of function spaces is to consider the diagonals of an upper triangular infinite matrix as an analogue of Fourier coefficients of an analytic function or distribution. Using this idea we consider a Bloch space of matrices (see [72]) and prove results similar to those which can be found in the well-known paper [3]. We mention the remarkable fact that this Bloch space is a commutative Banach algebra under the Schur product and that it is the topological dual of the Bergman-Schatten space.

Section 7.2 is dedicated to introduce and study the little Bloch space of matrices. In particular, it is proved that the dual space of the subspace of the little Bloch space consisting of compact matrices is the Bergman-Schatten space. (See Theorem 7.21.)

Chapter 8

Schur multipliers on analytic matrix spaces

An interesting topic in matriceal harmonic analysis is the study of Schur multipliers on different classes of Banach spaces of infinite matrices.

First we describe the Schur multipliers from $B(\ell_2)$ into $\mathcal{B}(D, \ell_2)$.

Theorem 8.1. *An upper triangular matrix A belongs to $(B(\ell_2), \mathcal{B}(D, \ell_2))$ if and only if $\sup_{0 \leq r < 1}(1-r)\|\sum_{k=1}^{\infty} kA_k r^{k-1}\|_{M(\ell_2)} < \infty$, or, equivalently, if and only if the $M(\ell_2)$-valued function $f_A(z) = \sum_{k=0}^{\infty} A_k z^k$ is a Lipschitz function with respect to the Bergman metric $\beta(z, w)$.*

Proof. Let $B \in B(\ell_2)$ and $A \in (B(\ell_2), \mathcal{B}(D, \ell_2))$ with $\|A\|_{(B(\ell_2), \mathcal{B}(D,\ell_2))} = C$. Then

$$\sup_{0 \leq r < 1} (1-r)\|\sum_{k=1}^{\infty} kA_k * B_k r^{k-1}\|_{B(\ell_2)} \leq C \cdot \|B\|_{B(\ell_2)}$$

and it follows that

$$\sup_{0 \leq r < 1} (1-r)\|\sum_{k=1}^{\infty} kA_k r^{k-1}\|_{M(\ell_2)} \leq C.$$

Conversely, if

$$\sup_{0 \leq r < 1} (1-r)\|\sum_{k=1}^{\infty} kA_k r^{k-1}\|_{M(\ell_2)} = C < \infty,$$

then, for an arbitrary $B \in B(\ell_2)$, we get obviously that

$$\sup_{0 \leq r < 1} (1-r)\|\sum_{k=1}^{\infty} kA_k * B_k r^{k-1}\|_{B(\ell_2)} \leq C \cdot \|B\|_{B(\ell_2)},$$

and, consequently,

$$A \in (B(\ell_2), \mathcal{B}(D, \ell_2)) \text{ and } \|A\|_{(B(\ell_2), \mathcal{B}(D,\ell_2))} \leq C.$$

175

On the other hand, it is clear that the proof of Theorem 5.1.6 in [94] holds also if we consider $M(\ell_2)$-analytic functions instead of usual analytic ones. Thus, $A \in (B(\ell_2), \mathcal{B}(D, \ell_2))$ if and only if the function $f_A(z)$ is a Lipschitz function with respect to the Bergman metric. The proof is complete.   □

Remark. The previous theorem may be regarded as a matrix extension of Theorem 4.3 of Jevtic and Pavlovic [47], namely:
  Theorem of Jevtic and Pavlovic. $(H^\infty, \mathcal{B}) = H_1^{1,\infty,1}$.

Next we consider the possibility to find the space of all Schur multipliers from $\mathcal{B}(D, \ell_2)$ into itself and expect that the following statement holds:

Theorem 8.2. $(\mathcal{B}(D, \ell_2), \mathcal{B}(D, \ell_2)) = H_1^{1,\infty,1}(\ell_2) = \{A \text{ upper triangular}$ *matrices such that* $\sup_{0<r<1}(1-r)\|A'(r)\|_{M(\ell_2)} < \infty\}$.

In order to do this we need an extension for matrices of a theorem of Shields and Williams [92].
Let $H_1^{1,\infty,1}$ be the space of all analytic functions g on D such that

$$\sup_{0<r<1}(1-r)\int_0^{2\pi}|g'(re^{i\theta})|\frac{d\theta}{2\pi} < \infty,$$

and $H_1^{\infty,\infty,1}$ the space of all analytic functions with

$$\sup_{0<r<1;\theta\in[0,2\pi)}(1-r)|g(re^{i\theta})| < \infty.$$

  Theorem of Shields and Williams: $(H^{\infty,\infty,1}, H^{\infty,\infty,1}) = H_1^{1,\infty,1}$. (See also Lemma 3.3 in [47].)

The matrix version of this last statement is the following:

Lemma 8.3. *It yields that*

$$H_1^{1,\infty,1}(\ell_2) \subset (H^{\infty,\infty,1}(\ell_2), H^{\infty,\infty,1}(\ell_2)),$$

where

$$H^{\infty,\infty,1}(\ell_2) = \{A \text{ upper triangular matrices such that}$$

$$\sup_{0<r<1}(1-r)\|A(r)\|_{B(\ell_2)} < \infty\}.$$

Proof. Let A be an upper triangular matrix such that

$$\sup_{r<1}(1-r)\|A'(r)\|_{M(\ell_2)} < \infty$$

and B such that

$$\sup_{r<1}(1-r)\|B(r)\|_{B(\ell_2)} < \infty.$$

We have to show that

$$\sup_{0<r<1}(1-r)\|(A*B)(r)\|_{B(\ell_2)} < \infty.$$

We note that

$$\sup_{0<r<1}(1-r)\|(A*B)(r)\|_{B(\ell_2)} \leq C \text{ if } \|(A'*B)(r)\|_{B(\ell_2)} \leq \frac{C}{(1-r)^2},$$

for all $0 < r < 1$.

Indeed, using Minkowski's inequality, we find that

$$\|(A*B)(\rho)\|_{B(\ell_2)} = \|\sum_{k=1}^{\infty}\frac{(A'*B)_k}{k}\rho^k\|_{B(\ell_2)} \leq$$

$$K\int_0^1\|\sum_k(A'*B)_k r^k\rho^k\|_{B(\ell_2)}dr \leq K\int_0^1\frac{1}{(1-r\rho)^2}dr = \frac{C}{1-\rho}.$$

The proof is complete. □

Proof of Theorem 8.2. It is clear that $(H^{\infty,\infty,1}(\ell_2), H^{\infty,\infty,1}(\ell_2)) \subset (\mathcal{B}(D,\ell_2), \mathcal{B}(D,\ell_2))$.

Consequently, using Lemma 8.3 and Theorem 8.1, we get that

$$(\mathcal{B}(D,\ell_2), \mathcal{B}(D,\ell_2)) \subset (B(\ell_2), \mathcal{B}(D,\ell_2)) = H^{1,\infty,1}(\ell_2)$$

$$\subset (H^{\infty,\infty,1}(\ell_2), H^{\infty,\infty,1}(\ell_2)) \subset (\mathcal{B}(D,\ell_2), \mathcal{B}(D,\ell_2)). \qquad \square$$

Theorem 8.4. *It yields that*

$$(L_a^1(D,\ell_2), L_a^1(D,\ell_2)) = H_1^{1,\infty,1}(\ell_2)$$

$$:= \{A \text{ upper triangular matrices such that } \sup_{0<r<1}(1-r)\|A'(r)\|_{M(\ell_2)} < \infty\}.$$

Proof. We use the obvious fact that the space (X, X) equals the space (X', X'), where X is a Banach space of infinite matrices and X' is the topological dual of X.

Then the theorem follows from Theorem 8.2. □

Next we want to present an application of Bloch matrices to describing the Schur multipliers from a matriceal version of the Hardy space to a matrix version of the BMO space, extending a nice result of Mateljevic and Pavlovic [60].

First we give a formula for the norm in $B(\ell_2)$ due to V. Lie [50] (see also Chapter 2):

$$||B||_{B(\ell_2)} = \sup_{||h||_{L^2(0,1)} \leq 1} \left(\sum_{k=1}^{\infty} | \int_0^1 \mathcal{L}_k^B(x)h(x)dx|^2 \right)^{1/2}.$$

Moreover, we use this formula to prove an interesting inequality also due to V. Lie.

Proposition 8.5. *If* $A \in H^1(\ell_2)$ *and* $B \in \mathcal{B}(D, \ell_2)$, *then we have that*

$$\left(\int_0^1 ||(A * B)'(r)||_{B(\ell_2)}^2 (1 - r)dr \right)^{1/2} \leq C||A||_{H^1(\ell_2)}||B||_{\mathcal{B}(D,\ell_2)},$$

where $C > 0$ *is a constant independent of the choice of* A *and* B.

Proof. From the above formula applied to $(A * B)'$ we may find a fixed $h \in L^2([0,1])$ with $||h||_2 = 1$ such that

$$||(A * B)'(r)||_{B(\ell_2)}^2 \sim \sum_{j=1}^{\infty} | \int_0^1 \left(\widetilde{\mathcal{L}}_j^A * \widetilde{\mathcal{L}}_j^B \right)' (re^{2\pi it})e^{2\pi ijt}h(e^{2\pi it})dt|^2.$$

We use now the following relation, which holds if f and g are analytic functions on the unit disk D and if $z = qe^{2\pi i\zeta}$, where $1 \geq |z| = q > 0$:

$$z(f * g)'(z) = 2 \int_0^{\sqrt{q}} \int_0^1 f'(re^{2\pi i(\theta+\zeta)})g'(re^{-2\pi i\theta})re^{2\pi i\zeta}d\theta dr,$$

and we get, by duality, that there is a $x = (x_j)_{j\geq 1} \in \ell_2(\mathbb{N})$, with $||x||_2 \leq 1$, such that

$$||(A * B)'(r)||_{B(\ell_2)} \approx | \sum_{j=1}^{\infty} x_j \int_0^1 \left(\widetilde{\mathcal{L}}_j^A * \widetilde{\mathcal{L}}_j^B \right)' (re^{2\pi it})e^{2\pi ijt}h(e^{2\pi it})dt| \sim$$

$$| \sum_{j=1}^{\infty} \int_0^{\sqrt{r}} \int_0^1 (\widetilde{\mathcal{L}}_j^A)'(se^{2\pi i\theta})x_j(\int_0^1 (\widetilde{\mathcal{L}}_j^B)'(se^{-2\pi i(\theta-t)})e^{2\pi ijt}h(e^{2\pi it})dt)sdsd\theta|$$

$$\leq \int_0^{\sqrt{r}} \int_0^1 \left(\sum_{j=1}^{\infty} |x_j|^2 | \left(\widetilde{\mathcal{L}}_j^A \right)' (se^{2\pi i\theta})|^2 \right)^{1/2} \times$$

$$\left(\sum_{j=1}^{\infty} | \int_0^1 (\widetilde{\mathcal{L}}_j^B)'(se^{-2\pi i(\theta-t)})e^{2\pi ijt}h(e^{2\pi it})dt|^2 \right)^{1/2} sdsd\theta \leq$$

$$C||B||_{\mathcal{B}(D,\ell_2)} \times \int_0^{\sqrt{r}} \frac{\int_0^1 \left(\sum_{j=1}^{\infty} |x_j|^2 |(\widetilde{\mathcal{L}}_j^A)'(se^{2\pi i\theta})|^2 \right)^{1/2} d\theta}{1-s} ds.$$

From this relation we deduce that

$$\int_0^1 ||(A*B)'(r)||_{\mathcal{B}(\ell_2)}^2 (1-r)dr \leq C||B||_{\mathcal{B}(D,\ell_2)}^2 \times$$

$$\sup_{||x||_2 \leq 1} \int_0^1 (1-r) \left(\int_0^{\sqrt{r}} \frac{\int_0^1 (\sum_{j=1}^{\infty} |x_j|^2 |(\widetilde{\mathcal{L}}_j^A)'(se^{2\pi i\theta})|^2)^{1/2}d\theta}{1-s} ds \right)^2 dr.$$

Using now Proposition 4.4 we find that

$$\int_0^1 ||(A*B)'(r)||_{\mathcal{B}(\ell_2)}^2 (1-r)dr \leq C||B||_{\mathcal{B}(D,\ell_2)}^2 ||A||_{H^1(\ell_2)}^2.$$

The proof is complete. $\qquad\qquad\qquad\qquad\qquad\qquad\qquad\qquad\square$

Now we prove the extension of the result of Mateljevic and Pavlovic:

Theorem 8.6. *We have the following relation:*

$$\mathcal{B}(D,\ell_2) = \big(H^1(\ell_2), BMOA(\ell_2) \big)$$

with equivalence of norms.

Proof. First we prove that

$$\big(H^1(\ell_2), BMOA(\ell_2) \big) \supset \mathcal{B}(D,\ell_2).$$

It is enough to show that

$$||A*B||_{BMOA(\ell_2)} \leq K||A||_{H^1(\ell_2)}||B||_{\mathcal{B}(D,\ell_2)},$$

for all $A \in H^1(\ell_2)$, and $B \in \mathcal{B}(D,\ell_2)$, where $K > 0$ is a constant not depending on A, that is, according to Corollary 5.4, we have to prove that:

$$\left(\int_0^1 ||(A*B)||_{\mathcal{B}(\ell_2)}^2 (1-r)dr \right)^{1/2} \leq C||A||_{H^1(\ell_2)}||B||_{\mathcal{B}(D,\ell_2)},$$

where $C > 0$ is a constant independent of the choice of A and B. But this is just Proposition 8.5.

The opposite inclusion is easy. We have to prove that

$$(H^1(\ell_2), BMOA(\ell_2)) \subset \mathcal{B}(D, \ell_2).$$

In order to do this, we assume that the following relation holds:

$$(H^1(\ell_2), \mathcal{B}(D, \ell_2)) \subset \mathcal{B}(D, \ell_2). \tag{8.1}$$

Then it is enough to prove that: $BMOA(\ell_2) \subset \mathcal{B}(D, \ell_2)$.

We remark that, for an upper triangular matrix A, we have that

$$||A||_{\mathcal{L}^2(\ell_2)} = \sup_k ||\mathcal{L}_k^A||_{L^2(\mathbb{T})}, \tag{8.2}$$

where $\mathbb{T}$ is the unidimensional torus.

With the same notations as above, by [18], we have that

$$||A||_{B(\ell_2)} = \sup_{||h||_2 \leq 1} \left(\sum_{k=1}^{\infty} | \int_0^1 \mathcal{L}_k^A(s)h(s)ds|^2 \right)^{1/2} \leq \sup_k ||\mathcal{L}_k^A||_{L^\infty(\mathbb{T})},$$

which, in its turn, by [33] and (8.2), implies that

$$||A||_{\mathcal{B}(D,\ell_2)} \leq \sup_k ||\mathcal{L}_k^A||_{\mathcal{B}(D)} \leq \sup_k ||\mathcal{L}_k^A||_{BMOA} \leq ||A||_{BMOA(\ell_2)}.$$

Hence, we have to prove only (8.1). This is the assertion of the next Proposition so the proof is complete when this result is proved. □

Proposition 8.7. *We have that*

$$(H^1(\ell_2), \mathcal{B}(D, \ell_2)) \subset \mathcal{B}(D, \ell_2).$$

Proof.   Let B be an upper triangular matrix such that $A * B \in \mathcal{B}(D, \ell_2)$, for all $A \in H^1(\ell_2)$. By the closed graph theorem there is a constant $C > 0$ such that

$$||A * B||_{\mathcal{B}(D,\ell_2)} \leq C||A||_{H^1(\ell_2)}, \text{ for all } A \in H^1(\ell_2).$$

Now take the Toeplitz matrix A associated with function $\theta \to \frac{e^{i\theta}}{(1-re^{i\theta})^2}$. Thus,

$$||A||_{H^1(\ell_2)} = \frac{1}{2\pi} \int_0^{2\pi} \frac{d\theta}{|1 - re^{i\theta}|^2} = \frac{1}{1-r^2},$$

which in turn implies that

$$||A * B||_{\mathcal{B}(D,\ell_2)} \leq \frac{C}{1-r^2}.$$

On the other hand it is clear that $A * B = \sum_{k=0}^{\infty} k B_k r^{k-1} = B'(r)$, and, thus, $||B'(r)||_{\mathcal{B}(D,\ell_2)} \leq \frac{C}{1-r^2}$ for all $0 < r < 1$.

Moreover, using the definition of $\mathcal{B}(D, \ell_2)$, we have that

$$\|B * C'(r) * C'(\rho)\|_{B(\ell_2)} \leq \frac{C}{(1 - r^2)(1 - \rho^2)},$$

for all $0 < r, \rho < 1$, $C > 0$ depending only on B.

Letting $r = \rho$ we get that

$$\|B * C'(r) * C'(r)\|_{B(\ell_2)} \leq \frac{C}{(1 - r^2)^2}, \quad \forall\, 0 < r < 1. \tag{8.3}$$

Moreover,

$$\int_0^r [B * C'(s) * C'(s)]ds = \int_0^r \sum_{k=0}^{\infty} k^2 B_k s^{2k-2}ds = \sum_{k=0}^{\infty} \frac{k^2}{2k-1} B_k r^{2k-1} =$$

$$r\left[\left(\sum_{k=0}^{\infty} B_k\right) * \left(\sum_{l=0}^{\infty} l r^{2l-1}\right) * \left(\sum_{n=1}^{\infty} \frac{n}{2n-1}\right)\right] = r[B * C'(r^2) * T],$$

where T is a Toeplitz matrix with $\frac{k}{2k-1}$ on the kth diagonal, $\forall\, k = 0, 1, 2, \ldots$

Now let T_1 be the Toeplitz matrix having on the diagonal $(T_1)_k$, $k = \pm 1, \pm 2, \pm 3, \ldots$, the constant sequence $\frac{2|k| \mp 1}{|k|}$ and on the main diagonal $(T_1)_0$ the value 2. It follows that

$$r\|B * C'(r^2) * T\|_{B(\ell_2)} \leq \int_0^r \|B * C'(s) * C'(s)\|_{B(\ell_2)}ds \leq \text{(by (8.3))} \leq$$

$$\int_0^r \frac{C}{(1-s^2)^2}ds \leq C_1 \int_0^r \frac{ds}{(1-s^2)^2} = C_1 \frac{r}{1-r} \leq C_2 \frac{1}{1-r^2},$$

where C_2 is an universal constant.

Therefore,

$$\|B * C'(r^2)\|_{B(\ell_2)} \leq \|B * C'(r^2) * T\|_{B(\ell_2)} \cdot \|T_1\|_{M(\ell_2)} \leq C_2 \|T_1\|_{M(\ell_2)} \leq$$

$$\frac{C_4}{1-r^2} \quad \forall\, 0 < r < 1,$$

that is, $B \in \mathcal{B}(D, \ell_2)$. The proof is complete. $\qquad\square$

Remark 8.8. *The result above complement the results of O. Blasco [12], [13]. In these papers Blasco introduced a space $\mathcal{B}(X)$ of X-valued Bloch functions, X a Banach space, which in the case $X = B(\ell_2)$, is larger than our space $\mathcal{B}(D, \ell_2)$, and working with another (more general) definition of multipliers space (X, Y) he proved that $(H^1(X), BMOA(X)) = \mathcal{B}(\ell_2)$ if and only if X is a Hilbert space. We also have to mention that if X is the scalar field of complex numbers, then the result of Mateljevic and Pavlovic follows also from Blasco's theorem.*

The following result extends Corollary 1 in [60]:

Corollary 8.9. *It yields that*
$$(H^1(\ell_2), VMOA(\ell_2)) = \mathcal{B}(D, \ell_2).$$

Proof.   Let $A \in H^1(\ell_2)$ and $B \in \mathcal{B}(D, \ell_2)$. Then, by Theorem 8.6,
$$||X * B||_{BMOA(\ell_2)} \le c||X||_{H^1(\ell_2)} \text{ for all } X \in H^1(\ell_2), \qquad (8.4)$$
where the constant c does not depend on X. If we substitute $X = A(s) - A$ in (8.4), we get that
$$||A * B(s) - A * B||_{BMOA(\ell_2)} \le c||A(s) - A||_{H^1(\ell_2)}.$$
Since obviously the term on the right-hand side of the last inequality approaches 0 when $s \to 1$, it follows that $A * B \in VMOA(\ell_2)$. The proof is complete. $\square$

Let $n_1, n_2 < \dots$ be a lacunary sequence of integers in the sense that
$$n_{k+1}/n_k \ge q > 1.$$
It is well-known that $g(z) = \sum_{k=1}^{\infty} z^{n_k} \in \mathcal{B}$, (see for instance the more general Theorem 7.9), so that we have by the previous corollary:

Corollary 8.10. *If $A = \sum A_n \in H^1(\ell_2)$, then, for every lacunary sequence $\{n_k\}$,*
$$\tilde{A} := \sum_{k=1}^{\infty} A_{n_k} \in VMOA(\ell_2).$$

Remark. It follows by the definition of $H^1(\ell_2)$ that the space of all upper triangular Hilbert-Schmidt matrices T_2 is contained in $H^1(\ell_2)$, so that if $A \in T_2$, then we have that $\sum_{k=1}^{\infty} A_{2^k} \in VMOA(\ell_2)$.

Since clearly $VMOA(\ell_2) \subset BMOA(\ell_2) \subset H^2(\ell_2)$, we get the matrix version of Paley's theorem:

Theorem 8.11. *If $A = \sum A_n \in H^1(\ell_2)$, then $P(A) := \sum_{k=1}^{\infty} A_{2^k} \in H^2(\ell_2)$.*

Another consequence of Corollary 8.10 is the following extension of a remark of [60]:

Corollary 8.12. *It yields that $A = \sum_{k=1}^{\infty} A_{2^k} \in VMOA(\ell_2)$ if and only if $A \in H^2(\ell_2)$, that is, if and only if*
$$\sup_k \sum_{j=1}^{\infty} |a_{2^j}^k|^2 < \infty,$$
where $A = (a_{2^j}^k)_{j,k \ge 1}$.

Proof. Since, clearly, $H^2(\ell_2) \subset H^1(\ell_2)$ the assertion of Corollary 8.12 follows from Theorem 8.11 and the above discussion. □

We mention also the extension of Corollary 2 in [60], which follows by Theorems 7.16 and the proof (see relation (8.1)) of Theorem 8.6:

Corollary 8.13. *The space of all Schur multipliers* $(H^1(\ell_2), \mathcal{B}_0(D, \ell_2))$ *coincides with* $\mathcal{B}(D, \ell_2)$.

Notes

Various results about Schur multipliers on different analytic matrix spaces are given in this chapter. For instance in Theorem 8.1 it is showed that
$$A \in (B(\ell_2), \mathcal{B}(D, \ell_2)) \text{ if and only if } \sup_{0 \le r < 1} (1 - r)\|\sum_{k=1}^{\infty} kA_k r^{k-1}\|_{M(\ell_2)} < \infty.$$
Another important result is Theorem 8.2, which gives a characterization of the space $(\mathcal{B}(D, \ell_2), \mathcal{B}(D, \ell_2))$.

But the most important result of this section is an answer to the problem to find a matrix version of the beautiful theorem of Matjelevic and Pavlovic [60] about the space of all Fourier multipliers from H^1 into $BMOA$. This is Theorem 8.6 mainly due to V. Lie and its proof uses results from Chapters 4 and 5. The statement is:
$$\mathcal{B}(D, \ell_2) = (H^1(\ell_2), BMOA(\ell_2)).$$

As one of its consequences we mention only a matrix version of Paley's theorem:

Theorem 8.11 If $A = \sum_{k=1}^{\infty} A_{2^k} \in H^1(\ell_2)$, then $P(A) = \sum_{k=1}^{\infty} A_{2^k} \in H^2(\ell_2)$.

Bibliography

[1] J. Arazy, *Some remarks on interpolation theorems and the boundness of the triangular projection in unitary matrix spaces,* Int. Eq. Oper. Th. **1**, (1978), 453-495.

[2] J. Arazy, *On the geometry of the unit ball of unitary matrix spaces,* Int. Eq. Oper. Th., **4** (1981), 151-171.

[3] J. M. Anderson, J. Clunie and Ch. Pommerenke, *On Bloch functions and normal functions,* J. Reine Angew. Math.,**270** (1974), 12-37.

[4] J. M. Anderson and A. Shields, *Coefficient multipliers of Bloch functions,* Trans. Amer. Math. Soc., **224** (1976), 255-265.

[5] J. Arazy, S. D. Fisher and J. Peetre, *Möbius invariant function spaces,* J. Reine Angew. Math. **363**(1985), 110-145.

[6] A. B. Alexandrov, Essays on non locally convex Hardy classes, Lecture Notes in Math. **864** (1981), 1-89, Springer-Verlag, Berlin-Heidelberg-New York.

[7] A. B. Alexandrov, *On the boundary decay in the mean of harmonic functions,* St. Petersburg Math. J. **5** (1996), 507-542.

[8] A. B. Alexandrov and V. V. Peller, *Hankel and Toeplitz-Schur multipliers,* Math. Ann., **324** (2002), 277-327.

[9] W. Arveson, *Subalgebras of C^*- algebras,* Acta Math. **123** (1969), 144-224.

[10] W. Arveson, *Interpolation problems in nest algebras,* J. Funct. Analysis **20** (1975), 208-233.

[11] G. Bennett, *Schur multipliers,* Duke Math. J., **44** (1977), 603-639.

[12] O. Blasco, *A characterization of Hilbert spaces in terms of multipliers between spaces of vector-valued analytic functions,* Michigan Math. J., **42**(1995), 537-543.

[13] O. Blasco, *Vector-valued analytic functions of bounded mean oscillation and geometry of Banach spaces,* Illinois J. Math., **41** (1997), 532-558.

[14] O. Blasco, *Introduction to vector valued Bergman spaces,* Function spaces and operator theory, 9–30, Univ. Joensuu Dept. Math. Rep. Ser., 8, Univ. Joensuu, Joensuu, 2005.

[15] O. Blasco and A. Pelczynski, Theorems of Hardy and Paley for vector valued analytic functions and related classes of Banach spaces, Trans. Amer. Math. Soc. 323 (1991), 335-367.

[16] M. Bozejko, *Littlewood functions, Hankel multipliers and power bounded operators on a Hilbert space,* Colloquium Math. **51** (1987), 35-42.

[17] S. Barza, L. E. Persson and N. Popa, *A matriceal analogue of Fejer's theory,* Math. Nachr.,**260** (2003), 14-20.

[18] S. Barza, V. D. Lie and N. Popa, *Approximation of infinite matrices by matriceal Haar polynomials,*Ark. Mat. **43** (2005), no. 2, 251–269.

[19] L. Bergh and J. Lofstrom, *Interpolation spaces. An Introduction,* Springer Verlag, Berlin, 1976.

[20] S. Barza, D. Kravvaritis and N. Popa, *Matriceal Lebesgue spaces and Hoelder inequality,* J. Funct. Spaces Appl. **3** (2005), no. 3, 239–249.

[21] R. R. Coifman, *A real variable characterization of H^p,* Studia Math., **51** (1974), 269-274.

[22] R. G. Cooke, Infinite matrices and sequence spaces, Macmillan 1950.

[23] P. L. Duren, Theory of H^p Spaces, Academic Press, 1970.

[24] R. G. Douglas, Banach algebra techniques in operator theory, Academic Press, New York, 1972.

[25] M. Déchamps-Godim, F. Lust-Piquard and H. Queffelec, *On the minorant properties in $C_p(H)$,* Pacific J. Math. **119** (1985), 89-101.

[26] P. Dodds and F. Sukochev, *RUC-decompositions in symmetric operator spaces,* Integr. Equat. Oper. Th., **29**(1997), 269-287.

[27] P. L. Duren, B. W. Romberg and A. L. Shields, *Linear functionals on H^p spaces with $0 < p < 1$,* J. Reine Angew. Math., **238** (1969), 32-60.

[28] P. L. Duren and A. Schuster, Bergman spaces, American Mathematical Society, Mathematical Surveys and Monographs, vol. 100, Providence, 2004.

[29] P. L. Duren and A. L. Shields, *Properties of H^p $(0 < p < 1)$ and its containing Banach space,* Trans. Amer. Math. Soc., **141** (1969), 255-262.

[30] R. E. Edwards, Functional Analysis; Theory and Applications, Holt, Rinehart and Winston, New-York, 1965.

[31] L. Fejér, *Untersuchungen über Fouriersche Reihen,* Math. Ann., **58** (1904), 501-569.

[32] C. L. Fefferman and E. M. Stein, *H^p spaces of several variables,* Acta Math., **129**(1972), 137-193.

[33] J. B. Garnett, Bounded analytic functions, Academic Press, New York, 1981.

[34] I. C. Gohberg and M. G. Krein, *Introduction to the theory of linear nonselfadjoint operators* Translated from the Russian by A. Feinstein. Translations of Mathematical Monographs, Vol. 18 American Mathematical Society, Providence, R.I. 1969

[35] A. Grothendieck, *Resumé de la théorie métrique des produits tensoriels topologiques,* Bol. Soc. Mat. Sao Paulo, **8** (1956), 1-79.

[36] W. W. Hastings, *A Carleson measure theorem for Bergman spaces,* Proc. Amer. Math. Soc. **52** (1975), 237-241.

[37] A. Haar, *Zur Theorie der orthogonalen Funktionensysteme,* Math. Ann., **69**, (1910), 331-371.

[38] A. Harcharras, *Fourier analysis, Schur multipliers on S^p and noncommutative $\Lambda(p)$-sets,* Studia Math., **137** ,(3), (1999), 203-260.

[39] G. H. Hardy and J. E. Littlewood, *Notes on the theory of series (XX) Gen-*

eralizations of a theorem of Paley, Quart. J. Math., Oxford Ser., **8** (1937), 161-171.

[40] G. H. Hardy and J. E. Littlewood, *A new proof of a theorem on rearrangements,* J. London Math. Soc. **23** (1948), 163-168.

[41] K. Hoffman, Banach Spaces of Analytic Functions, Prentice Hall, Englewood Cliffs, 1962.

[42] J. R. Holub, *On the metric geometry of Ideals of Operators on Hilbert Space,* Math. Ann. **201** (1973), 157-163.

[43] C. Horowitz, *Zeros of functions in the Bergman spaces,* Duke Math. J. **41**(1974), 693-710.

[44] C. Horowitz, *Factorization theorems for functions in the Bergman spaces,* Duke Math. J. **44** (1977), 201-213.

[45] U. Haagerup and G. Pisier, *Factorization of analytic functions with values in non-commutative L_1−spaces and applications,* Can. J. Math.,**XLI**, (1989), 882-906.

[46] D. R Jocic and D. Krtinic, *Schur-Laurent multipliers for block matrices and geometric characterization of continuous matrices* , Linear and Multilinear Algebra, vol. **58**,(2010), 523-534.

[47] M. Jevtic and M. Pavlovic, *Coefficient multipliers on spaces of analytic functions,* Acta Sci. Math. (Szeged), **64** (1998), 531-545.

[48] D. Krtinic, *A matricial analogue of Fejer's theory for different types of convergence,* Math. Nachr., **280** (2007), 1-6.

[49] S. Kwapien and A. Pelczynski, *The main triangle projection in matrix spaces and its applications,* Studia Math.,**34**,(1970), 43-68.

[50] V. Lie, *Integral operators in infinite matrix theory; the study of matriceal Riesz projection,* (Romanian), Master's dissertation at University of Bucharest, 49p, 2003.

[51] F. Lust-Piquard, *Inégalités de Khintchine dans C_p $(1 < p < \infty)$,* C. R. Acad. Sc. Paris, **303** , Série **I**, (1986), 289-292.

[52] F. Lust-Piquard, *On the coefficient problem: a version of the Kahane Katznelson de Leeuw theorem for spaces of matrices,* J. Funct. Anal., **149** (1997), 352-376.

[53] F. Lust-Piquard and G. Pisier, *Non-commutative Khintchine and Paley inequalities,* Ark. Mat. (29)(1991), 241-260.

[54] J. Lindenstrauss and H. P. Rosenthal, *The $\mathcal{L}_p$ spaces,* Israel J. Math. **7** (1969), 325-349.

[55] A. Marcoci, *Some new results concerning Lorentz sequence spaces and Schur multipliers,* Licentiate thesis, Luleå University of Technology, Sweden, 2009.

[56] L. Marcoci, *Some new results concerning Schur multipliers and duality results between Bergman-Schatten and little Bloch spaces,* Licentiate thesis, Luleå University of Technology, Sweden, 2009.

[57] A. Marcoci, *Some new results concerning Banach spaces of infinite matrices and Lorentz sequence spaces,* Doctoral thesis, Luleå University of Technology, Sweden, 2010.

[58] L. Marcoci, *A study of Schur multipliers and some Banach spaces of infinite*

matrices, Doctoral thesis, Luleå University of Technology, Sweden, 2010.

[59] M. Marsalli, *Noncommutative H^2 spaces*, Proc. Amer. Math. Soc., **125** (1997), 779-784.

[60] M. Mateljevic and M. Pavlovic, *Multipliers of H^p and $BMOA$*, Pacific J. Math. **146** (1990), 71-84.

[61] M. Mateljevic and M. Pavlovic, L^p- *behaviour of the integral means of analytic functions,* Studia Math. **77**, (1984), 219-237.

[62] L.G. Marcoci, L.E. Persson, I. Popa and N. Popa, *A new characterization of BergmanSchatten spaces and a duality result,* J. Math. Anal. Appl, **360** (2009), 67-80.

[63] N. Nikolskii, *Treatise on the shift operator,* Springer Verlag, Berlin, 1986.

[64] D. M. Oberlin, *Translation-invariant operators on $L^p(G)$,* $0 < p < 1$, Michigan Math. J. **23** (1976), 119-122.

[65] V. Peller, *Hankel Operators and Their Applications* , Springer- Verlag, New York, 2003.

[66] V. Peller, *A description of Hankel operators of class S_p for $p > 0$, an investigation of the rate of rational approximation and other applications,* Mat. Sbornik, **122** (1983), 481-510.

[67] V. Paulsen, *Completely bounded maps and dilations,* Pitman Research Notes in Math. 146, Longman, Wiley, New-York, 1986.

[68] M. Pavlović, *Introduction to function spaces on the disk.* Matematicki Institut SANU, Beograd, 2004.

[69] L. B. Page, *Bounded and compact vectorial Hankel operators,* Trans. Amer. Math. Soc., **150** (1970), 529-539.

[70] S. Parrott, *On a quotient norm and the Sz.-Nagy-Foiaş lifting theorem,* J. Funct. Anal., **30** (1978), 311-328.

[71] J. Peetre, *New thoughts on Besov spaces,* Duke Univ. Press., Durham, N.C., 1976.

[72] N. Popa, *Matriceal Bloch and Bergman-Schatten spaces,* Rev. Roumaine Math. Pures Appl., **52** (2007), 459-478.

[73] N. Popa, *A characterization of upper triangular trace class matrices,* C. R. Acad. Sci. Paris, Ser. I 347 (2009), 59-62.

[74] N. Popa, *Schur multipliers between Banach spaces of matrices,* Proceedings of the Sixth Congress of Romanian Mathematicians Bucharest, 2007, vol. 1, 373-380

[75] I. Popa and N. Popa *Matrices of bounded variation ,* in Proceeding of the International Conference: Mathematical Analysis and its Applications, Athens 2002, pp. 318-323.

[76] I. Popa and N. Popa *Inequalities in matrix spaces,* Math. Rep., **12(62)** (2010), 169-180.

[77] S. Power, *Commutators with the triangular projection and Hankel forms on nest algebras,* J. London Math. Soc. 32 (1985), 272-282.

[78] I. I. Privalov, Randeigenschaften analytischer Funktionen, Verlag der Wiss., Berlin, 1956.

[79] A. Pelczynski and F. Sukochev, *Some remarks on Toeplitz multipliers and*

Hankel matrices, Studia Math., **175**, (2006), 175-204.

[80] G. Pisier, *Multipliers of the Hardy space H^1 and power bounded operators,* Preprint, 2000.

[81] G. Pisier, *Similarity problems and completely bounded maps,* LNM 1618, Springer Verlag, Berlin, 1995.

[82] A. Pietsch, *Operator Ideals,* VEB Deutscher Verlag der Wissenschaften, Berlin, 1978.

[83] A. Pietsch and H. Triebel, *Interpolationstheorie für Banachideale von beschränkten linearen Operatoren,* Studia Math., **31**,(1968), 95-109.

[84] A. L. Shields, *An analogue of a Hardy-Littlewood-Fejer inequality for upper triangular trace class operators,* Math. Z., **182** (1983), 473-484.

[85] D. Sarason, *Generalized interpolation in H^∞,* Trans. Amer. Math. Soc., **127** (1967), 179-203.

[86] I. Schur, *Bemerkungen zur Theorie der beschränkten Bilinearformen mit unendlich vielen Veränderlichen,* J. Reine Angew. Math., **140** (1911), 1-28.

[87] B. Simon, *Trace ideals and their applications,* London Math. Soc Lecture Notes Series, no. 35, Cambridge, 1979.

[88] W. T. Sledd and D. A. Stegenga, An H^1 multiplier theorem, Ark. Mat., **19** (1981), no. 2, 265-270.

[89] B. Smith, *A strong convergence theorem for $H^1(\mathbb{T})$,* Banach spaces, harmonic analysis and probability theory, (Storrs, Conn., 1980/1981), Lecture Notes in Math., vol. 995, Springer-Verlag, Berlin, 1983, 169-173.

[90] E. M. Stein and R. Shakarchi, Fourier analysis; An Introduction, Princeton University Press, Princeton, 2003.

[91] W. Stinespring, *Positive functions on C^*-algebras,* Proc. Amer. Math. Soc., **6** (1966), 211-216.

[92] A. L. Shields and D. L. Williams, *Bounded projections, duality and multipliers in spaces of analytic functions,* Trans. Amer. Math. Soc., **162** (1971), 287-302.

[93] H. Triebel, *Interpolation theory, function spaces, differential operators.* Second edition. Johann Ambrosius Barth, Heidelberg, 1995.

[94] K. Zhu, Operator theory in Banach function spaces, Marcel Dekker, New-York, 1990.

[95] K. Zhu, Analytic Besov space, J. Math. Anal. Appl., **157**, (1991), 318-336.

[96] A. Zygmund, Trigonometric series, Cambridge University Press, Cambridge, 1959.

[97] N. Wiener, *The quadratic variation of a function and its Fourier coefficients,* Massachusett's J. Math., **3** (1924), 72-94.

[98] G. Wittstock, *Ein Operatorenwertigen Hahn-Banach Satz,* J. Funct. Anal., **40** (1981), 127-150.

Index

$(H^1 - \ell_1)$-multiplier, 80
(X, Y) the space of all Schur multipliers from X into Y, 1
$(\cdot, \cdot)$ the scalar product in a Hilbert space, 3
(p, q)-bounded, 80
$A * B$ the Schur product of the matrices A and B, 1
A_k the kth-diagonal matrix , 1
$B(\ell_2)$ the space of all bounded linear operators on ℓ_2, 2
$BMOA$ the space of analytic functions of bounded mean oscillation, 4
$BMOA(\ell_2)$, 109
$BMO_F(\ell_2)$, $BMOA_F(\ell_2)$, xi, 114
C_1 the trace class, 3
C_2 the Hilbert-Schmidt class, 3
C_p, $0 < p < \infty$ Schatten class, 3
D_A the defect operator, 92
E_t the Toeplitz matrix with entries $(e^{i(j-k)t})_{j,k \geq 1}$, 71
FM, 84
$H^2(\mathcal{H})$ Hardy space of Hilbert space-valued functions, 112
$H^{\infty,\infty,1}$, $H^{\infty,\infty,1}(\ell_2)$, 176
$H^\infty(\ell_2)$, 114
$H^p(\ell_2)$ the matriceal Hardy space of index p, 67
H^p, $0 < p \leq \infty$ Hardy spaces, 4
H^p_X vector-valued Hardy space, 79
$H_1^{1,\infty,1}(\ell_2)$, 176

H_Φ matrix version of Hankel operator, 113
$L^1(D, \ell_2)$, $L^\infty(D, \ell_2)$, 5
$L_a^{p,unc}(D, \ell_2)$, 132
$M(\ell_2)$ the space of all Schur multipliers on $B(\ell_2)$, 2
$M(\mathbb{T})$ the convolution algebra of Borel measures on $\mathbb{T}$, 101
P, $\widetilde{P}$ Bergman projection, 128
P_T the triangular projection, 69
P_α, 129
$T_p(\ell_R^2)$, $Tp(\ell_C^2)$ spaces of upper triangular matrices, 70
T_p, $0 < p < \infty$ the matrix analogue of Hardy space H^p, 69
$VMO_F(\ell_2)$, 118
Γ_Ω the block Hankel matrix, 96
$\beta(z, w)$ the Bergman metric, 7
$\ell_\infty(\ell_\infty)$, $\ell_2(\ell_2, w)$, 134
$\mathcal{B}$ Bloch space, 4
$\mathcal{B}(D, \ell_2)$ the matriceal Bloch space, 150
$\mathcal{B}_{0,c}(D, \ell_2)$, 170
$\mathcal{B}_0$ the little Bloch space, 4
$\mathcal{B}_0(D, \ell_2)$ little Bloch space, 161
$\mathcal{C}(\overline{D}, \ell^2)$, 165
$\mathcal{C}_0(D, C_\infty)$, 171
$\mathcal{C}_0(D, \ell_2)$, 169
$\mathcal{C}_0 \widehat{\otimes}_\epsilon C_\infty$ ϵ-tensor product of corresponding Banach spaces, 171
$\mathcal{I}$, 152
$\mathcal{L}_k^A$, $\widetilde{\mathcal{L}}_k^A$, $\mathcal{L}_A(x, t)$, 65

$\mathcal{M}^{p,2}$, 119

$\mathcal{T}$, 2

$\rho(z, w)$ the pseudo-hyperbolic
   distance, 7

$\sigma_n(A)$, 137

$f_A(r, t) = \sum_{k=-\infty}^{\infty} A_k(r)e^{ikt}$, 2

m_X the operator induced by the
   multiplier m, 80

w^*-measurable function, 5

$VMOA(\ell_2)$, 109

Bergman-Schatten classes, 122

Analytic matrices, 2

Banach space of $(H^1 - \ell_1)$-Fourier
   type, 81

Bennett's Theorem, 2

Bloch matrix, 150

crudely finitely representable Banach
   space, 86

Matriceal Hausdorff-Young Theorem,
   75

Matrix version of Nehari theorem, xi,
   114

Pavlović Theorem, 97

Schmidt Theorem, 3

Shields inequality, 70

singular values of an operator T, 3

strongly measurable function, 5

the Hankel matrix $\mathcal{H}_a$, 90

The matrix spaces $\mathcal{L}_r^p(\ell_2)$, resp.
   $\mathcal{L}_c^p(\ell_2)$, $1 \leq p \leq 2$, 66

The noncommutative factorization
   theorem, 87

the projective tensor product, 88

Toeplitz matrices, 2

Vectorial Nehari theorem, 113